Lecture Notes in Mathematics

A collection of informal reports and seminars
Edited by A. Dold, Heidelberg and B. Eckmann, Zürich

309

David H. Sattinger
University of Minnesota, Minneapolis, MN/USA

Topics in Stability and Bifurcation Theory

Springer-Verlag
Berlin · Heidelberg · New York 1973

AMS Subject Classifications (1970): 35-02, 35B35, 35G20, 35J60, 35K55, 35Q10, 46-xx, 76D05, 76E99

ISBN 3-540-06133-9 Springer-Verlag Berlin · Heidelberg · New York
ISBN 0-387-06133-9 Springer-Verlag New York · Heidelberg · Berlin

Offsetdruck: Julius Beltz, Hemsbach/Bergstr.

Preface

In analyzing the dynamics of a physical system governed by nonlinear equations the following questions present themselves: Are there equilibrium states of the system? How many are there? Are they stable or unstable? What happens as external parameters are varied? As the parameters are varied, a given equilibrium may lose its stability (although it may continue to exist as a mathematical solution of the problem) and other equilibria or time periodic oscillations may branch off. Thus, bifurcation is a phenomenon closely related to the loss of stability in nonlinear physical systems.

The subjects of bifurcation and stability have always attracted the interest of pure mathematicians, beginning at least with Poincaré and Lyapounov. In the past decade an increasing amount of attention has focused on problems in partial differential equations. The purpose of these notes is to present some of the basic mathematical methods which have developed during this period. They are primarily mathematical in their approach, but it is hoped they will be of value to those applied mathematicians and engineers interested in learning the mathematical techniques of the subject.

In a number of sections we have explained the basic mathematical tools needed for the development of the subject of bifurcation theory, for example elements of the theory of elliptic boundary value problems, the Riesz-Schauder theory of compact operators, the Leray-Schauder topological degree theory, and the implicit function theorem in a Banach space. Nevertheless, the reader will have to have a certain amount of background in mathematical analysis, particularly in the areas of partial differential equations and functional analysis, in order to benefit from these notes.

These notes were the basis of a course given at the University of Minnesota during the academic year 1971-1972. The author would like to express his thanks to Professors D. G. Aronson, Gene Fabes, Charles McCarthy, and Daniel Joseph for their lively interest in the course, and for their stimulating and cogent remarks. Thanks are also due Miss L. Ruppert and Miss P. Williams for their fine job of typing. He would also like to acknowledge the support of the Air Force contract in the production of these notes. (AFOSR 883-67)

We hope these notes prove a convenient source of references and stimulate further interest in a diverse subject.

D. H. Sattinger
Minneapolis
May 10, 1972

Contents

Introduction

1.1 Ordinary Differential Equations

In this course we are going to consider the related problems of stability, instability, and bifurcation theory for simple nonlinear parabolic problems, and for the Navier Stokes equations. Bifurcation is a phenomenon peculiar to nonlinear problems and is closely related to the loss of stability. In developing the subject, we shall have the opportunity to consider a number of mathematical techniques, and so the course should prove valuable in attaining a certain degree of mathematical proficiency - particularly in regard to maximum principles, energy and variational principles, topological methods (fixed point theory and Leray-Schauder degree theory), and functional analysis.

Let us begin by reviewing some simple facts and examples from ordinary differential equations. These provide a guide for what to look for in the case of partial differential equations, although the techniques of partial differential equations are often quite different.

Consider a system of ordinary differential equations

$$\dot{x} = f(x,\mu) \tag{1.1.1}$$

where $x = (x_1, \dots x_n)$, $f = (f_1, \dots f_n)$, and μ is a parameter.

Suppose $f(x_0,\mu_0) = 0$ for some point x_0 in R^n , $\mu = \mu_0$; then x_0 is called an equilibrium solution.

<u>Definition 1.1.1</u>. <u>We say that x_0 is a stable equilibrium of (1.1.1) if, given any $\epsilon > 0$ there is a $\delta > 0$ such that $|v - x_0| < \delta$ implies $|x(t) - x_0| < \epsilon$ for all $t > 0$, where $x(t)$ is the solution of (1.1.1) with initial data v. We say that x_0 is asymptotically stable if in addition $|x(t) - x_0| \to 0$ as $t \to \infty$.</u>

Here $|x| = \sqrt{x_1^2 + \ldots + x_n^2}$ or any equivalent norm.

<u>Theorem 1.1.2</u>: (Lyapounov) Let $A = \left.\left(\frac{\partial f_i}{\partial x_j}\right)\right|_{(x_0,\mu_0)}$ <u>be the Jacobian of f at (x_0,μ_0). If all eigenvalues of A have negative real parts then x_0 is a stable equilibrium. If some eigenvalues of A have positive real parts, then x_0 is unstable.</u>

Theorem 1.1.2 can be called "the principle of linearized stability." The reason for this is the following.

Let the solution of (1.1.1) be written in the form $x(t) = u(t) + x_0$; thus u is the perturbation from equilibrium. From $\dot{x} = f(x,\mu_0)$ we get

$$\dot{u} = \dot{x} = f(u + x_0,\mu_0) = f(x_0,\mu_0) + \frac{\partial f_i}{\partial x_j} u_j + O(|u|^2) ,$$

$$\dot{u} = Au + O(|u|^2) \tag{1.1.2}$$

where $A = \left.\frac{\partial f_i}{\partial x_j}\right|_{x_0,\mu_0}$. $O(|u|^2)$ denotes a term $g(u)$ such that $|g(u)| \leq c|u|^2$. If we neglect the second order term in (1.1.2) we obtain the linear equation

$$\dot{u} = Au \tag{1.1.3}$$

whose solution is $u(t) = e^{tA}u_0$. As is well known, all solutions of (1.1.3) decay provided the spectrum of A lies in the left half plane; some solutions of (1.1.3) grow exponentially if A has eigenvalues in the right half plane.

It is argued that the second order term may be neglected when the perturbations are small. This heuristic reasoning is justified by Lyapounov's theorem. In the case of partial differential equations we shall prove theorems of an analogous nature. The term $O(|u|^2)$ appearing in (1.1.2) may in that case, however, be a nonlinear partial differential operator (for example, $u \cdot \nabla u$ in the case of the Navier Stokes equations) and the extension of Lyapounov's theorem to this case is not immediately obvious.

At any rate, we call Lyapounov's theorem the principle of linearized stability because it says it is enough to look at the linearized problem in order to determine stability. Incidentally, if some of the eigenvalues of $\frac{\partial f_i}{\partial x_j}$ have zero real parts (critical case) then it is not enough to look at the linearized problem - and one must look at the effects of the nonlinear terms in order to determine stability or instability.

Now let us consider the case where $x_0(\mu)$ is an equilibrium solution for μ in an interval $a \leq \mu \leq b$:

$$f(\mu, x_0(\mu)) = 0 .$$

Suppose that as μ crosses μ_0 , $a < \mu_0 < b$, some of the eigenvalues of $A(\mu) = \frac{\partial f_i}{\partial x_j}(\mu, x_0(\mu))$ cross the imaginary axis, as in the diagrams below:

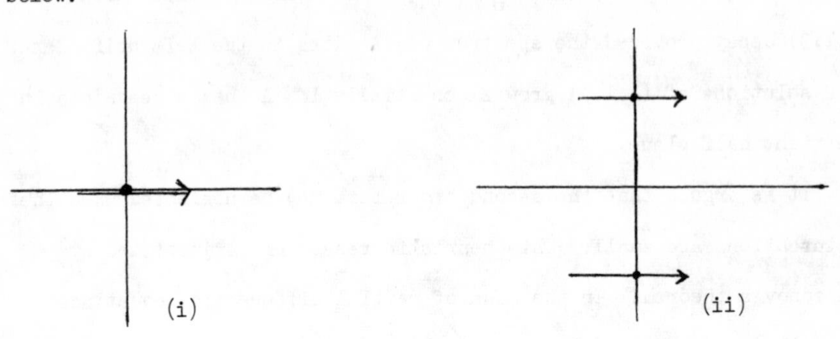

Thus, for $\mu < \mu_0$, $x_0(\mu)$ is stable, while $x(\mu)$ becomes unstable when $\mu > \mu_0$. What happens? This is where the phenomenon of bifurcation occurs.

Two important cases are those in which a simple eigenvalue crosses through the origin (i) or in which a pair of simple complex conjugate eigenvalues cross the imaginary axis (ii). (We are restricting ourselves here to the case where $f(x,\mu)$ is real for real x and μ ; this includes a wide range of physical problems. In that case the eigenvalues of $A(\mu)$ always appear in complex conjugate pairs.) In these two cases one can give a fairly complete description of what happens.

In both cases we shall assume

$$\operatorname{Re} a'(\mu_0) > 0 ,$$

where $a(\mu)$ is one of the critical eigenvalues of $A(\mu)$ (the simple real one in case (i), one of the complex ones in case (ii)). Then

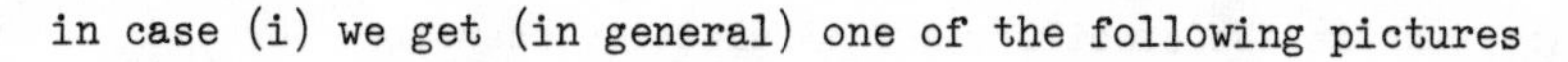

in case (i) we get (in general) one of the following pictures

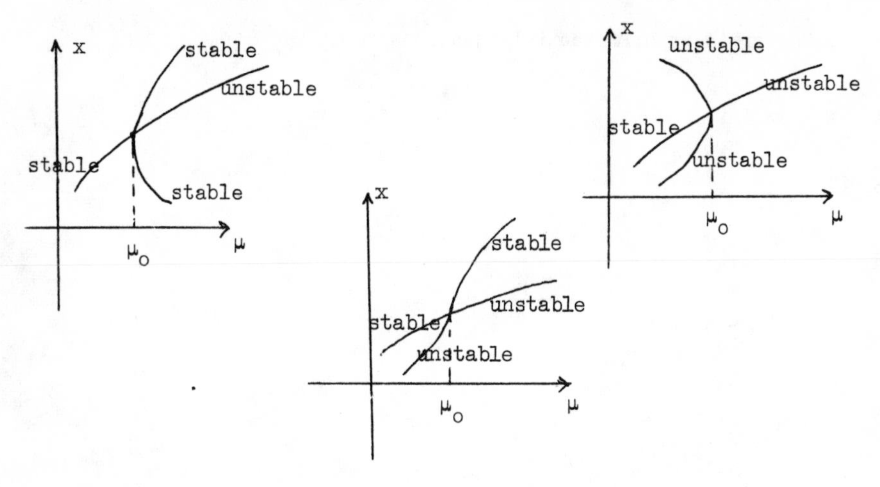

Each curve in the diagrams represents a solution curve $(\mu, x(\mu))$ of the equations $f(x(\mu),\mu) = 0$. The point $(\mu_0, x_0(\mu))$ is a <u>bifurcation point</u> - two solution branches intersect. The labels "stable" and "unstable" denote which solution branches are stable equilibria of the problem $f(\mu,x(\mu)) = 0$. We see that the given branch $x_0(\mu)$ is stable for $\mu < \mu_0$ and unstable for $\mu > \mu_0$, as we assumed. In the case of the bifurcating branch, solutions which bifurcate above criticality are stable while solutions which bifurcate below criticality are unstable.

In case (ii) we get no bifurcation of stationary solutions, but a bifurcation of time periodic motions of $\dot{x} = f(x,\mu)$. As opposed to the stationary case, we get only one periodic motion — (Note that there were two in the stationary case) — apart from phase shifts. Again, solutions which bifurcate above criticality are stable while solutions which bifurcate below criticality are unstable.

Exercises: Discuss the stability and bifurcation phenomena for the following ordinary differential equations

1. $\dot{x} = \mu x + x^2$

2. $\dot{x} = \mu x - x^2$

3. $\dot{x} = \mu x + x^3$

4. $\dot{x} = \mu x - x^3$

5. $\dot{x} = \mu x - y + x(x^2+y^2)$

   $\dot{y} = \mu y + x + y(x^2+y^2)$

6. $\dot{x} = \mu x - y - x(x^2+y^2)$

   $\dot{y} = \mu y + x - y(x^2+y^2)$

Discuss the stability both of the trivial solutions and the bifurcating solutions.

7. $\dot{x} = \mu x - y + p(x,y)$ where p,q are analytic and

   $\dot{y} = \mu y + x + q(x,y)$ $p = q = O(x^2+y^2)$ as $x,y \to 0$.

1.2 Examples of Bifurcation in Physical Systems

Physical systems which exhibit the phenomenon of bifurcation range, literally, from yardsticks to stars. The buckling of a yardstick when the thrust on the end is increased past a critical value is a familiar example of bifurcation. Below the critical thrust the yardstick maintains a vertical position which is stable to displacements. Above the critical thrust the yardstick bows to one side or the other, and the vertical, unbowed, position is no longer stable. Astrophysicists have attempted to explain the formation of binary stars by a process of loss of stability and bifurcation. This problem was the subject of classic papers by Poincaré (1885) and Lyapounov (1908); the ideas of both of these mathematicians concerning bifurcation continue to this day to have a strong influence on the subject. Below we discuss a few specific physical systems which exhibit bifurcation phenomena and mention briefly some of the mathematical models used to describe these systems.

Chemical Kinetics

The general equations governing chemical reactions between various agents are

$$\frac{\partial C_i}{\partial t} = k_i \Delta C_i + f_i(C_1, \ldots C_n, T)$$

$$\frac{\partial T}{\partial t} = K\Delta T + g(C_1, \ldots C_n, T) \tag{1.2.1}$$

where $C_1, \ldots C_n$ are the concentrations, and T is the temperature. Equations (1.2.1. are supplemented by boundary conditions, such as

$$\frac{\partial T}{\partial \nu} + \beta T = 0 \qquad \text{or} \qquad T = T_0$$

$$\frac{\partial C_i}{\partial \nu} = 0 \qquad \text{or} \qquad C_i = C_{i,0} \tag{1.2.2}$$

Equations (1.2.1) - (1.2.2) form a parabolic boundary value problem.

Their derivation and specific forms for the reaction rates g_i and f may be found in the book by Gavalas.

A special simple case which is often considered in thermal combustion processes is

$$\frac{\partial C}{\partial t} = K\Delta C - ce^{-E/RT}$$
$$\frac{\partial T}{\partial t} = k\Delta T + QCe^{-E/RT} \tag{1.2.3}$$

The rate factor $\exp\{-E/RT\}$ is called the Arrenhius rate factor; Q is the heat of reaction, E is called the activation energy, and R is the gas constant. If, as is sometimes done, it is assumed that the concentration remains constant during the initial stages of the reaction, we get the following equation for the temperature

$$\frac{\partial T}{\partial t} = k\Delta T + QC_0 e^{-E/RT} \quad ,$$

where C_0 is the initial concentration. Steady state solutions of this equation are governed by

$$0 = \Delta T + \lambda e^{-E/RT} \quad , \quad \lambda = \frac{QC_0}{k} \tag{1.2.4}$$

Numerical analysis (see Parks) of this problem on a sphere of radius r_0 has shown the existence of multiple solutions (with boundary conditions $T = T_0$ on ∂D). By plotting $T(0)$, the center temperature, against λ one obtains a solution diagram with roughly the shape shown

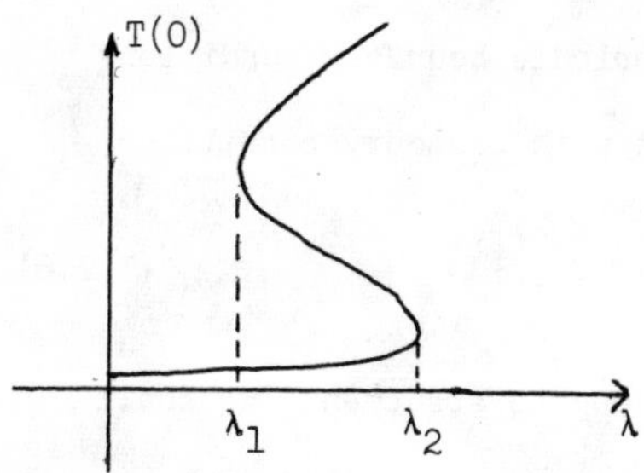

in the figure at the left. We see from the graph that there are three solutions for $\lambda_1 < \lambda < \lambda_2$ while for $\lambda < \lambda_1$ and $\lambda > \lambda_2$ there is only one solution. As λ increases past λ_1 two solutions suddenly appear. As λ increases one of these solutions increases while the other decreases until it meets the lower solution as λ reaches λ_2. The lower and middle solution then vanish as λ increases past λ_2. This phenomenon of multiple equilibrium solutions and their stability is of great interest in applications. Time periodic oscillations are also known to occur in other circumstances. (See Lee and Luss [2])

Fluid Dynamics

The motion of a visious incompressible fluid is governed by the Navier Stokes equations

$$\frac{\partial u_i}{\partial t} + u_j \frac{\partial u_i}{\partial x_j} = -\frac{\partial p}{\partial x_i} + \frac{1}{R} \Delta u_i$$

$$\frac{\partial u_i}{\partial x_i} = 0 \qquad (1.2.5)$$

Here u_1, u_2, u_3 are the Cartesian components of the fluid velocity, and p is the pressure. Repeated indices denote summation. The number R, called the Reynolds number, is a pure number which results from introducing non-dimensional variables. Equations (1.2.5) may be considered on infinite domains with periodic boundary conditions or they may be considered on a finite domain with boundary conditions

$$u_i \Big|_{\partial D} = \psi_i \tag{1.2.6}$$

In Chapter VII, using a topological degree arguement, we shall prove that equations (1.2.5) with boundary conditions (1.2.6) have a time independent solution for all values of R. For small values of R this solution is unique and unconditionally stable: all perturbations decay and the system returns to its equilibrium state. As R increases this basic solution loses its stability and new stationary solutions may branch off, or time periodic oscillations may set in.

In 1944, L. Landau proposed that turbulence arises from repeated loss of stability and branching. Thus, the secondary solution in turn loses its stability at higher values of R and is replaced by yet another stationary solution or perhaps by a time periodic solution. With the further increase in R successive bifurcations may take place, with periodic solutions replaced by quasi-periodic motions of more and more base periods. E. Hopf [2] constructed an intriguing mathematical model (a system of nonlinear equations which bears a similarity with the Navier Stokes equations) which exhibited precisely this behavior of repeated loss of stability and branching. Hopf then went on to construct a statistical mechanics for his model.

The situation as regarding fluid mechanics is, of course, somewhat more complicated. For example the bifurcation may be subcritical, and the bifurcating solutions unstable. This appears to be the case for plane Pouiseille flow (see Chapter VIII). It is also possible for a flow to lose stability with no bifurcation taking place at all. In such cases as these the fluid would not progress toward turbulence through a continuous branching process, but might make a sudden transition to turbulence when the critical parameter value is crossed. An interesting discussion of other matters related to Landau's ideas may be found in the article by Ruelle and Takens. They point out that true quasi-periodic motion is non-generic in the case of ordinary differential equations. Thus we may expect that turbulence should be more complicated than quasi-periodic motion.

Elasticity.

The stability and buckling of elastic systems subject to boundary and internal stresses is of prime interest in elasticity theory. The general equations of elasticity are quite complex, but much success has been had in treating simplified models. One such model is Von Kármán's equations for the buckling of a thin elastic plate:

$$\begin{aligned} \Delta \Delta f &= -[w,w] \\ \Delta \Delta w &= \lambda[F_0,w] + [f,w] \end{aligned} \tag{1.2.7}$$

where

$$[f,g] = f_{xx}g_{yy} + f_{yy}g_{xx} - 2f_{xy}g_{xy} .$$

Here w is the deflection of the plate and $F = \lambda F_0 + f$ is the stress. Equations (1.2.7) are taken together with the boundary conditions

$$f = f_x = f_y = 0$$

and

$$w = w_x = w_y = 0$$

for a clamped plate. The null solution $f = 0$, $w = 0$ represents the unbuckled state.

Notes

For a development of stability and bifurcation theory for ordinary differential equations see the books by Coddington and Levinson and J. Hale. The bifurcation of periodic solutions from an equilibrium solution was treated completely by E. Hopf [1] in 1942.

The theory of Chemical Kinetics is discussed in the books by Gavalas and Frank-Kamenetsky. Also see the article by Gelfand.

For recent articles on bifurcation problems in Stellar Structure, see S. Chandrasekhar [2] and N. Lebovitz.

A general treatment of Von Kármán's equations has been given in a series of papers by Berger and Fife. (The article in the bibliography contains a list of previous articles.) Von Kármán's equations on a circular plate, for radially symmetric buckling, has been considered by K. O. Friedrichs and J. J. Stoker, and by Keller, Keller, and Reiss.

A recent work which contains good discussions of a number of areas of interest in applications, as well as an extensive bibliography, is the review article by I. Stakgold.

II

Nonlinear Elliptic Boundary Value Problems of Second Order

2.1 Maximum Principles

The chief tool in the investigation of second order elliptic and parabolic problems is the maximum principle. An excellent reference for the subject is the book by Protter and Weinberger.

Let $x = (x_1, \dots x_n)$ and let a_{ij} be smooth functions of x with $\sum_{i,j=1}^{n} a_{ij}\xi_i\xi_j \geq \mu \sum_{i=1}^{n} \xi_i^2$ for some constant $\mu > 0$. This inequality is assumed to hold throughout a domain D. (A domain means an open connected set.) The operator L, defined by

$$L[u] = \sum_{i,j=1}^{n} a_{ij} \frac{\partial^2 u}{\partial x_i \partial x_j} + \sum_{i=1}^{n} b_i \frac{\partial u}{\partial x_i} ,$$

is then a uniformly elliptic operator.

Theorem 2.1.1: Let $Lu \geq 0$ in D. If $u \leq M$ in D and $u = M$ at an interior point of D, then $u \equiv M$ in D.

Corollary 2.1.2: If $Lu = 0$ in D then u attains its maximum and minimum values on ∂D.

Suppose now that $Lu \geq 0$ in D and that u attains its maximum at some point P on ∂D (the symbol ∂D denotes the boundary of D) It is clear that $\frac{\partial u}{\partial \nu} \geq 0$, where $\frac{\partial}{\partial \nu}$ denotes the outward normal derivative of u at P. However, the sharper statement $\frac{\partial u}{\partial \nu} > 0$ is also true: More precisely,

Theorem 2.1.3: (Maximum principle at the boundary) Let $Lu \geq 0$ in D, let $u \leq M$ in D and let $u = M$ at P on ∂D. Assume there is a sphere $K \subset D$ tangent to ∂D at P. If u is continuous at P and $\frac{\partial u}{\partial \nu}$ exists (where ν is any outward directional derivative) then $\frac{\partial u}{\partial \nu} > 0$ at P unless $u \equiv M$ in D. For proofs of Theorems 2.1.1 and 2.1.3, see pp. 61-68 of Protter and Weinberger's book.

To remember how the inequalities go it is useful to think of the one dimensional operator

$$Lu = u'' .$$

If $Lu \geq 0$ then u is convex and cannot attain a local maximum unless $u \equiv$ constant. Similarly, if $u'' \leq 0$ then u cannot attain a local minimum. For parabolic operators, that is, operators of the form

$$Lu - \frac{\partial u}{\partial t} ,$$

we also have a maximum principle:

Theorem 2.1.4: Let u satisfy

$$Lu - \frac{\partial u}{\partial t} \geq 0$$

in a domain $E = D \times (0,T)$, where D is a domain in R^n. If $u \leq M$ in E and $u = M$ at some point (x_1,t_1) with $x_1 \in D$ and $t_1 > 0$, then $u = M$ in $D \times (0,t_1]$.

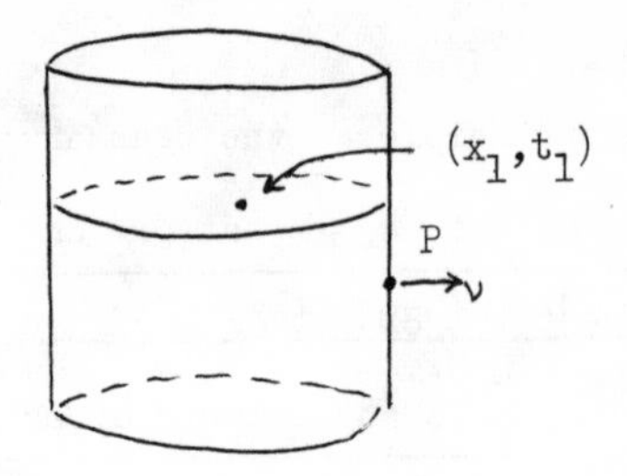

<u>Again we have a maximum principle at the boundary: If u assumes its maximum M at a point P on $\partial D \times (0,T)$ and if $u < M$ in the interior of $D \times (0,T)$, then $\frac{\partial u}{\partial \nu} > 0$ at P.</u>

<u>Exercises</u>:

1. (a) Use the maximum principle to prove uniqueness for the solutions of the boundary value problems

$$\Delta u = f \qquad u\big|_{\partial D} = g$$

and

$$\Delta u = f \qquad \frac{\partial u}{\partial \nu} + \beta u = g \quad (\beta \geq 0) \quad \text{on } \partial D .$$

(b) Does the maximum principle (Corollary 2.1.2) apply to the problem

$$\Delta u + u = 0 \qquad 0 \leq x,y \leq \pi$$

$$u\big|_{\partial D} = 0 \qquad ?$$

Why not? (Show it is not valid) What about the equation $\Delta u + c(x)u = 0$ where $c(x)$ is an arbitrary function?

2. Using the maximum principle, what can you say about solutions of the nonlinear problems

$$\Delta u = u^2 \qquad u\big|_{\partial D} = 0$$

$$\Delta u + \lambda u + u^3 = 0 \qquad u\big|_{\partial D} = 0 \ .$$

3. If $Lu + c(x,t)u - \frac{\partial u}{\partial t} \geq 0$ in a region $D \times (0,T)$ and if $u(x,0) \leq 0$ and $u(x,t) \leq 0$ for $x \in \partial D$ and $0 \leq t < T$, then $u(x,t) \leq 0$ in $D \times (0,T)$. Observe that it is not necessary to assume that $c(x,t)$ is negative to prove the result. (Compare with exercise 1b)

2.2 Elliptic Boundary Value Problems.

In the following sections we need some of the machinery of the theory of elliptic boundary value problems. For convenient reference, we summarize these results below as they apply to second order boundary value problems.

We first introduce the Hölder and Sobolev function classes. A function u is Hölder continuous with exponent α $(0 < \alpha \leq 1)$ on a domain $\bar{D}$ if $\sup|x-y|^{-\alpha}|u(x) - u(y)| < +\infty$ for all x,y in $\bar{D}$. The Hölder norm of u is defined by

$$|u|_\alpha = \sup_{x\in\bar{D}}|u(x)| + \sup_{x,y\in\bar{D}} \frac{|u(x) - u(y)|}{|x-y|^\alpha} \ .$$

Let $\ell = (\ell_1, \ldots \ell_n)$, where ℓ_i are nonnegative integers, and put

$$D^\ell u = \frac{\partial^{|\ell|} u}{\partial x_1^{\ell_1} \cdots \partial x_n^{\ell_n}}$$

where $|\ell| = \ell_1 + \ldots + \ell_n$. The class $C_{k+\alpha}$ consists of functions u which, together with their derivatives up to order k , are Hölder continuous with exponent α . The class $C_{k+\alpha}$ is a Banach space under the norm

$$|u|_{k+\alpha} = \sum_{|\ell|\leq k} |D^{\ell}u|_{\alpha} \quad .$$

The Sobolev classes $W_{k,p}$ consist of those functions u which have strong L_p derivatives up to order k ; these classes are Banach spaces under the norm

$$\|u\|_{k,p} = \left(\sum_{|\ell|\leq k} \int_D |D^{\ell}u|^p dx\right)^{1/p} \quad .$$

Theorem 2.2.1 (Embedding Lemma):

If D is a smoothly bounded domain in R^n and $n < p < \infty$ then each u in $W_{1,p}$ is equal almost everywhere to a function $\bar{u}$ in $C_{\alpha}(D)$ and

$$|\bar{u}|_{\alpha} \leq c(n,p,D) \; \|u\|_{1,p}$$

where $\alpha = 1 - n/p$ and the constant c is independent of u . (see Morrey's book, Theorem 3.6.6)

Now consider the boundary value problem

$$Lu = f \quad \text{in} \quad D \tag{2.2.1}$$

with the boundary conditions

$$Bu = a\frac{\partial u}{\partial \nu} + bu = g \tag{2.2.2}$$

We assume that $\nu = (\nu_1, \ldots \nu_n)$ is a smoothly varying $(C_{1+\alpha})$ outward normal vector field on ∂D, which is of class $C_{2+\alpha}$. The coefficients of L are assumed to be of class C_α while for the functions a and b we assume that either $b = 1$ and $a = 0$ or $a = 1$ and $b \in C_{1+\alpha}(\partial D)$. We assume $f \in C_\alpha$ and that g has an extension $\hat{g}$ to the interior of D such that $\hat{g} \in C_{2+\alpha}$.

Theorem 2.2.2: Under the above assumptions the boundary value problem (2.2.1) - (2.2.2) is uniquely solvable and satisfies the a-priori estimates

$$|u|_{2+\alpha} \leq c(|f|_\alpha + |\hat{g}|_{2+\alpha})$$

where the constant c is independent of f and g. The following L_p estimates of Agmon - Douglis - Nirenberg are also valid:

$$\|u\|_{2,p} \leq c(\|f\|_p + \|\hat{g}\|_{2,p}) \quad .$$

(See the paper of Agmon - Douglas - Nirenberg, especially Theorems 7.3 and 15.2)

For given f let Kf denote the solution of the boundary value problem (2.2.1) - (2.2.2) with homogeneous boundary data $(g=0)$. The transformation K is linear and takes L_p into $W_{2,p}$ for all $1 < p < \infty$. The nonlinear boundary value problem

$$\begin{aligned} Lu + f(x,u) &= 0 \\ Bu &= 0 \quad \text{on} \quad \partial D \end{aligned} \tag{2.2.3}$$

can be written

$$u + KF(u) = 0 \tag{2.2.4}$$

where $F(u)$ is the operator $u \to f(x,u)$. Using the a-priori estimates of Theorem (2.2.2) and the embedding lemma we can easily prove

Theorem 2.2.3: Let u be a bounded measurable function on D satisfying (2.2.4) and let f be Hölder continuous with exponent α . Then $u \in C_{2+\alpha}(\bar{D})$ and is a classical solution of (2.2.3).

The proof is left as an exercise.

The following result, known as the Fredholm alternative for elliptic operators, is of great importance.

Theorem 2.2.4: Consider the boundary value problem

$$\begin{aligned} Lu + c(x)u &= f \\ Bu &= 0 \end{aligned} \tag{2.2.5}$$

where c and f are of class C_α . Either problem (2.2.5) is boundedly invertible for arbitrary f or it has a nontrivial solution when $f = 0$. Theorem (2.2.4) is a consequence of the fact that the operator K defined above is compact and the Fredholm alternative for compact operators. The details of the proof will be indicated in exercises in Chapter III . Later, in Chapter IV , we shall discuss the solvability of (2.2.5) in the case where the homogeneous problem ($f = 0$) does have nontrivial solutions.

We finally want to consider eigenvalue problems of the type

$$\begin{aligned} L\varphi + c(x)\varphi + \lambda\varphi &= 0 \\ B\varphi &= 0 \end{aligned} \tag{2.2.6}$$

We assume either $b \geq 0$ in the boundary operator B if $a = 1$ or $b = 1$ and $a = 0$. For the time being, assume $c(x) < 0$ everywhere in D. We first note that the eigenvalue problem (2.2.6) may be written in the form

$$\varphi + \lambda A \varphi = 0$$

where A is the inverse of the operator $(L + c)$ with the boundary conditions $Bu = 0$. A is a compact operator (see Chapter III) on the Banach space $C(\bar{D})$ (continuous functions on $\bar{D}$ with the sup norm); furthermore, by the strong maximum principle, $-Af > 0$ in D if $f \geq 0$ on D.

A <u>cone</u> in a Banach space is a closed set of elements K with the properties that

(i) $u, v \in K$ imply $\alpha u + \beta v \in K$ for all $\alpha, \beta \geq 0$.

(ii) $u, v \in K$ and $u \neq 0$ implies that $u + v \neq 0$.

The set of non-negative functions is an example of a cone in the Banach space $C(\bar{D})$. The interior of this cone consists of strictly positive functions. If we take the Banach space $C_0(D)$ (continuous functions which vanish on ∂D) then the interior of K is the class of functions which are strictly positive in the interior of K.

An operator A is called <u>strongly positive</u> (relative to a cone K) if for each $u \in \partial K$ there is an integer $n = n(u)$ such that $A^n u$ belongs to the interior of K. In the present case, $-A = -(L + c)^{-1}$ is strongly positive with $n = 1$ for all u; this is an immediate consequence of the strong maximum principle. (We take $B = C_0(\bar{D})$ in case of the Dirichlet boundary conditions.) According to a theorem of Krein and Rutman (Theorem 6.3, p. 267 in their article) a strongly positive operator A has one and only one eigenfunction φ interior to the cone K and the corresponding eigenvalue is real and simple.

Therefore the eigenvalue problem (2.2.6) has one and only one positive eigenfunction, and the corresponding eigenvalue λ_1 is real and simple. In the case of the boundary conditions

$$\frac{\partial\varphi}{\partial\nu} + b\varphi = 0$$

where $b > 0$, the boundary point maximum principle shows that $\varphi > 0$ on ∂D. A further maximum principle argument (see pp. 89, 90 in Protter and Weinberger) shows that all other eigenvalues λ have real part strictly greater than λ_1.

Finally, in case $c(x)$ is not everywhere negative on D we replace c by $(c - \mu)$ and λ by $\lambda - \mu$ where μ is a constant, $\mu > c$. Then, repeating the arguments above, we see that the eigenvalue problem (2.2.6) still has a real simple eigenvalue of smallest real part (possibly negative) and that the corresponding eigenfunction is positive. We call these the principal eigenvalue and eigenfunction.

<u>Exercises</u>:

1. (a) Let φ_1 be the principal eigenfunction of

$$L\varphi + \lambda\varphi = 0$$

$$\varphi\big|_{\partial D'} = 0$$

where $\bar{D} \subset D'$. Show that $\varphi_1 > 0$ on $\bar{D}$.

(b) Let φ_1 be the principal eigenfunction for

$$L\varphi + \lambda\varphi = 0$$

$$\frac{\partial\varphi}{\partial\nu} + b\varphi = 0 \quad \text{on } \partial D .$$

where $b > 0$ on ∂D. Show that $\varphi_1 > 0$ on $\bar{D}$.

2. Let λ_1 be the principal eigenvalue of the Laplacian on a domain D. If $\mu < \lambda_1$ show there is a function φ satisfying

$$\Delta\varphi + \mu\varphi = 0 \qquad \varphi > 0 \quad \text{on } \bar{D}$$

Hint: Let $D' \supset \bar{D}$ and let φ be the principal eigenfunction of the Laplacian on D' with zero boundary conditions.

3. Let A be the operator

$$Au = \int_0^x u(t)dt .$$

Show that A is positive on $C[0,1]$ but not strongly positive. Does A have any eigenfunctions?

2.3 Monotone Iteration Schemes.

Consider the boundary value problem :

$$\begin{aligned} Lu + f(x,u) &= 0 \quad \text{in } D \\ Bu &= g \quad \text{on } \partial D , \end{aligned} \tag{2.3.1}$$

where B is the boundary operator in § 2.2 .

An upper solution to this problem is a function ϕ satisfying

$$L\phi + f(x,\phi) \leq 0 \quad \text{in } D ,$$

$$B\phi \geq g \quad \text{on } \partial D .$$

A <u>lower solution</u> is a function ψ satisfying

$$L\psi + f(x,\psi) \geq 0 \quad \text{in} \quad D$$

$$B\psi \leq g \quad \text{on} \quad \partial D \ .$$

We assume that ∂D, f, g, and the coefficients of L are smooth in what follows.

<u>Theorem 2.3.1</u>: <u>Let ϕ be an upper solution and ψ a lower solution, with $\psi \leq \phi$ on D. Then there exists a solution u to the boundary value problem (2.3.1) with $\psi \leq u \leq \phi$.</u>

<u>Proof</u>: Choose a number Ω such that

$$\frac{\partial f}{\partial u} + \Omega u > 0 \quad \text{on} \quad D \tag{2.3.2}$$

for $\min \psi \leq u \leq \max \phi$. We define a (nonlinear) transformation T by $v = Tu$ if

$$\begin{aligned} (L - \Omega)v &= -[f(x,u) + \Omega u] \quad \text{in} \quad D \\ Bv &= g \quad \text{on} \quad \partial D \ . \end{aligned} \tag{2.3.3}$$

Let us show that T is <u>monotone</u>: if $u \leq v$ then $Tu < Tv$. We have

$$(L - \Omega)Tu = -[f(x,u) + \Omega u]$$

$$(L - \Omega)Tv = -[f(x,v) + \Omega v]$$

and $Tu = Tv = g$ on ∂D. Putting $w = Tv - Tu$ we get

$$(L - \Omega)w = -[f(x,v) - f(x,u) + \Omega(v-u)] \ ,$$

$$Bw = 0 \quad \text{on} \quad \partial D$$

But the function $F(x,u) = f(x,u) + \Omega u$ is increasing in u by (2.3.2), so for $v \geq u$

$$0 \leq F(x,v) - F(x,u) = f(x,v) - f(x,u) + \Omega(v-u) .$$

Therefore

$$Lw \leq 0 \quad \text{in} \quad D$$

$$w = 0 \quad \text{on} \quad \partial D .$$

By the maximum principle, $w \geq 0$, so $Tu \leq Tv$.

Now let $u_1 = T\phi$. Let us prove that $u_1 \leq \phi$:

$$(L - \Omega)u_1 = -[f(x,\phi) + \Omega\phi]$$

$$Bu_1 = g \quad \text{on} \quad \partial D .$$

Therefore

$$(L - \Omega)(u_1 - \phi) = (L - \Omega)u_1 - (L - \Omega)\phi$$

$$= -[f(x,\phi) + \Omega\phi] - L\phi + \Omega\phi$$

$$= -[L\phi + f(x,\phi)] \geq 0$$

and $B(u_1 - \phi) = g - \phi \leq 0$ on ∂D .

By the maximum principle $(u_1 - \phi) \leq 0$ in D: $u_1 = T\phi \leq \phi$.

Now define $u_2 = Tu_1$. $u_1 < \phi \Longrightarrow u_2 = Tu_1 \leq T\phi = u_1$. By induction we get a monotone decreasing sequence of iterates

$$\phi \geq u_1 \geq u_2 \geq \cdots$$

Similarly, starting from ψ we get a monotone increasing sequence $\psi \leq v_1 \leq v_2 \leq \cdots$ where $v_1 = T\psi$ and $v_n = Tv_{n-1}$.

Moreover, $\psi < \phi \Longrightarrow T\psi < T\phi$, i.e. $v_1 < u_1$. By induction $v_n < u_n$ for all n:

$$\psi \leq v_1 \leq v_2 \leq \cdots \quad \leq u_2 \leq u_1 \leq \phi .$$

Therefore both sequences converge (since they are monotone). Let $\bar{v} = \lim_n v_n$, $\bar{u} = \lim_n u_n$ and we have $\bar{v} \leq \bar{u}$. Let us show that $\bar{v}$ and $\bar{u}$ are <u>fixed points</u> of T: $\bar{v} = T\bar{v}$, and $\bar{u} = T\bar{u}$.

In fact, from the L_p estimates of Theorem (2.2.2) and the continuity of f we see that T takes bounded pointwise convergent sequences into sequences which converge in $W_{2,p}$ for $1 < p < \infty$. By the embedding lemma (taking $p > n$) we may then conclude that the iterations converge in C_α for $\alpha = 1 - n/p$. Returning to theorem (2.2.2) and applying the Schauder estimates we see that the iterations converge in $C_{2+\alpha}$. Therefore

$$\bar{u} = \lim_n u_n = \lim_n Tu_{n-1} = T(\lim_n u_{n-1}) = T(\bar{u}) \quad .$$

Since $\bar{u}$ is in $C_{2+\alpha}$ it is a classical solution of the boundary value problem.

One can apply similar arguments to parabolic initial value problems:

$$Lu + f(x,u) - \frac{\partial u}{\partial t} = 0$$

$$u|_{\partial D} = g(x,t) \quad , \quad u(x,0) = u_0(x) \qquad (2.3.4)$$

$$x \in D \quad , \quad 0 \leq t \leq T$$

Call $\phi(x,t)$ an upper solution of (2.3.4) if

$$L\phi + f(x,\phi) - \frac{\partial\phi}{\partial t} \leq 0 ,$$

$$\phi(x,t)|_{\partial D} \geq g(x,t) \quad , \quad \phi(x,0) \geq u_0(x) \quad .$$

A lower solution ψ is defined by reversing these inequalities. By arguments similar to those in the proof of Theorem 2.3.. we can prove the following.

Theorem 2.3.2: Let there exist upper and lower solutions ϕ and ψ for (2.3.4), with $\psi < \phi$ in $D \times (0,T)$. Then there exists a solution $u(x,t)$ of (2.3.4) satisfying $\psi(x,t) \leq u(x,t) \leq \phi(x,t)$.

The proof of Theorem (2.3.2) is left as an exercise.

Exercises:

1. Let the plane E^2 be partially ordered by $x < y$ if $x_1 < y_1$ and $x_2 < y_2$, where $x = (x_1,x_2)$ and $y = (y_1,y_2)$. Call T a monotone operator on E_2 if $x < y$ implies $Tx < Ty$. Show if there exist points a,b such that $a < b$, $a < Ta$ and $Tb < b$, then there exists a fixed point c, $a < c < b$ and $c = Tc$.

2. In the problem above suppose there exist points a,b such that $a < b$, $Ta < a$ and $Tb > b$. Does T have a fixed point c such that $a < c < b$?

3. (Perron): (a) if $\dot{x} < g(t,x)$ and $\dot{y} = g(t,y)$, $x(0) \leq y(0)$, then $x(t) < y(t)$ for $t > 0$.

   (b) if $\dot{x} \leq g(t,x)$ and $\dot{y} = g(t,y)$, $x(0) \leq y(0)$, then $x(t) \leq y(t)$ for $t > 0$. Hint: let y_ϵ satisfy $\dot{y}_\epsilon = g(t,y_\epsilon) + \epsilon$ for $\epsilon > 0$ and apply (a) .

(c) Let $\frac{\partial g}{\partial x} \geq 0$ and let $x \geq g(t,x)$, $y \leq g(t,y)$ with $x(0) \leq y(0)$. Show there exists a solution of $\dot{z} = g(t,z)$ with $x(0) \leq z(0) \leq y(0)$ and that $x(t) \leq z(t) \leq y(t)$.

4. (a) If $x(t)$ (scalar) satisfies the integral inequality

$$x(t) \leq a(t) + \int_0^t f(s,x)ds \qquad t \geq 0$$

where f is increasing in x, then $x(t) \leq y(t)$, where y satisfies

$$y(t) = a(t) + \int_0^t f(s,y)ds$$

Use part (a) to estimate $x(t)$ if

$$x(t) \leq 1 + \int_0^t x^2(s)ds \quad .$$

5. Prove Theorem 2.3.2.

2.4 <u>A Simple Bifurcation Problem</u>.

Consider the boundary value problem

$$\begin{aligned} (\Delta + \mu)u - u^3 &= 0 \quad \text{in } D , \\ u|_{\partial D} &= 0 \quad . \end{aligned} \tag{2.4.1}$$

Let λ_1 be the principle eigenvalue of the Laplacian:

$$\Delta \varphi_1 + \lambda_1 \varphi_1 = 0 \qquad\qquad \varphi_1|_{\partial D} = 0 \quad .$$

<u>Theorem 2.4.1</u>: <u>The problem (2.4.1) has no nontrivial solutions for $\mu < \lambda_1$.</u>

<u>Proof</u>: The eigenvalues of the Laplacian depend monotonically and continuously on the domain (with mild qualifications on the variation

of the domain). See, for example, Courant and Hilbert, vol. I, p. 423. Specifically, if $\mu < \lambda_1$ there is a domain $D' \supset D$ such that

$$\Delta \psi_1 + \mu \psi_1 = 0 \quad ,$$

$$\psi_1 \Big|_{\partial D'} = 0 \quad .$$

Then ψ_1 is strictly positive on $\bar{D}$. If u is any solution of (2.4.1) we put $u = w\psi_1$ and substitute into (2.4.1) to get the following equation for w :

$$\psi_1 \Delta w + 2(\nabla\psi_1)\cdot(\nabla w) + w\Delta\psi_1$$
$$+ \mu w \psi_1 - w^3 \psi_1^3 = 0 \quad .$$

Since $\psi_1 > 0$ on $\bar{D}$ we can divide through by ψ_1 to get

$$\Delta w + \frac{2}{\psi_1}(\nabla\psi_1)\cdot\nabla w + w(-w^2\psi_1^2) = 0 \ .$$

The coefficient of w is negative so by the maximum principle $w \equiv 0$ ($w = 0$ on ∂D) and $u = 0$ is therefore the only solution of the boundary value problem for $\mu < \lambda_1$.

Theorem 2.4.2: For $\mu > \lambda_1$ equation (2.4.1) has at least two nontrivial solutions —— one positive and one negative.

Proof: We construct upper and lower solutions. First we try $u = \sigma\varphi_1$ where σ is a constant and φ_1 is the principle eigenfunction of the Laplacian. We have

$$(\Delta + \mu)\sigma\varphi_1 - \sigma^3\varphi_1^3$$
$$= \sigma\varphi_1((\mu - \lambda_1) - \sigma^2\varphi_1^2) \quad .$$

For $\mu > \lambda_1$ the quantity in parentheses is positive when $|\sigma|$ is small, so $\sigma\varphi_1$ is a lower solution when $\sigma > 0$ and $|\sigma|$ is small; and $\sigma\varphi_1$ is an upper solution when $\sigma < 0$ and $|\sigma|$ is small.

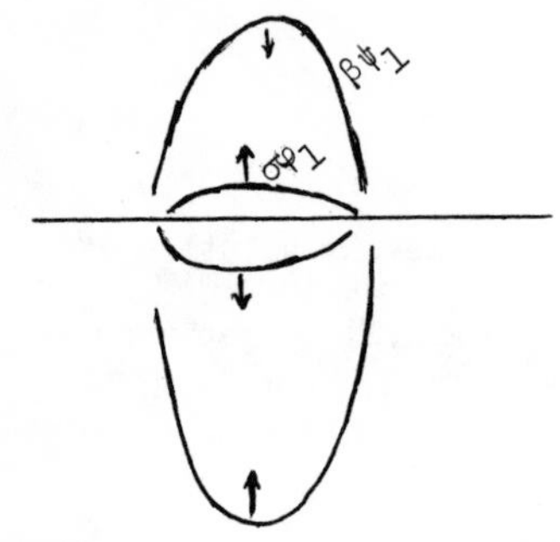

To get a positive upper solution we try $\beta\psi_1$ where ψ_1 is the principal eigenfunction of

$$\Delta\psi_1 + \lambda_1'\psi_1 = 0$$

$$\psi_1 = 0 \quad \text{on } D' ,$$

Then we get

$$(\Delta + \mu)\beta\psi_1 - (\beta\psi_1)^3 = ((\mu - \lambda_1') - \beta^2\psi_1^2)\beta\psi_1 \quad .$$

Since $\psi_1 > 0$ on $\bar{D}$ the quantity in parentheses above is negative for suitably large β . Therefore $\beta\psi_1$ is an upper solution for large positive β and a lower solution for large negative β .

Theorems (2.4.1) and (2.4.2) taken together show that two nontrivial solutions of (2.4.1) bifurcate from the trivial solution when μ increases past λ_1 . The situation may be represented schematically by the "bifurcation" diagram below.

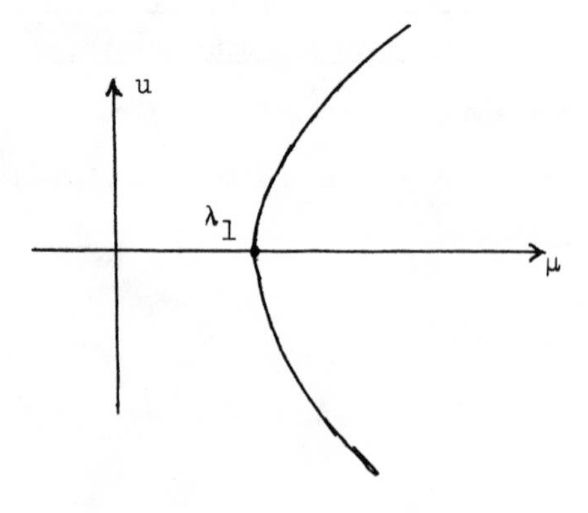

We note further: If φ_1 is normalized so that $\sup \varphi_1 = 1$ and $0 \leq \varphi_1 \leq 1$, then for our lower solutions $\sigma\varphi_1$ we may take σ as large as we please, provided that

$$(\mu - \lambda_1) - \sigma^2\varphi_1^2 \geq 0 \quad .$$

Since $\varphi_1 \leq 1$ we may take $\sigma = \sqrt{\mu - \lambda_1}$. This shows that the positive nontrivial solution lies above $\sqrt{\mu - \lambda_1}\ \varphi_1$ and tends to infinity with μ.

Now we wish to examine the stability properties of the various solutions. We first prove the stability of the null solution when $\mu < \lambda_1$. Let $u(x,t)$ satisfy

$$\begin{aligned} &(\Delta + \mu)u - u^3 - \frac{\partial u}{\partial t} = 0 \\ &u\Big|_{\partial D} = 0 \qquad u(x,0) = u_0(x) \ . \end{aligned} \tag{2.4.2}$$

We shall prove:

<u>Theorem 2.4.3</u>: <u>When $\mu < \lambda_1$ there exist constants $K > 0$ and $\sigma > 0$ (depending only on $\lambda_1 - \mu$) such that the solution u of (2.4.2) satisfies</u>

$$|u(x,t)| \leq Ke^{-\sigma t} \sup|u_0(x)| \quad .$$

<u>Proof</u>: Let ψ_1 be any function satisfying $\Delta\psi_1 + \lambda_1'\psi_1 = 0$, $\psi_1 > 0$ on $\bar{D}$, where $\mu < \lambda_1' < \lambda_1$. Normalize ψ_1 so that $\sup|\psi_1| = 1$ and put $u(x,t) = w\psi_1 e^{-\sigma t}$, with $\sigma > 0$ to be chosen later. We get for w the equation

$$\Delta w + 2\frac{\nabla\psi_1}{\psi_1}\cdot\nabla w + w\,((\mu - \lambda_1') - w^2\psi_1^2 + \sigma) - \frac{\partial w}{\partial t} = 0$$

$$w\Big|_{\partial D} = 0 \qquad w(x,0) = \frac{u_0(x)}{\psi_1(x)} .$$

If σ is chosen so that $\mu - \lambda_1' + \sigma \leq 0$ then the quantity in braces is always negative, so by the maximum principle

$$|w(x,t)| \leq \sup |w(x,0)| = \sup \frac{|u_0(x)|}{|\psi_1(x)|} .$$

The theorem now follows by taking $K = \sup\frac{1}{|\psi_1(x)|}$. Q.E.D.

In the next section we develop some machinery for stability investigations of the nontrivial solutions.

<u>Exercises</u>:

1. Try investigating the bifurcation problem for

   (a) $\Delta u + \mu u + u^3 = 0$

   (b) $\Delta u + \mu u + \mu^2 = 0$

by the methods of this section. What do you find? (hint: the methods won't work, at least not those given by theorem 2.4.1. But what would you guess to be the case?)

2. Investigate the possibility of multiple solutions of the boundary value problem

$$\Delta T + \lambda e^{-E/RT} = 0 ,$$

$$T\Big|_{\partial D} = T_0 .$$

2.5 Initial Value Problems.

Theorem 2.5.1: Let ϕ be an upper solution:

$$L\phi + f(x,\phi) \leq 0 \ , \qquad \phi \geq g(x) \quad \text{on} \quad \partial D$$

and let u satisfy

$$\begin{aligned} & Lu + f(x,u) - \frac{\partial u}{\partial t} = 0 \\ & u\Big|_{\partial D} = g(x) \qquad\qquad u(x,0) = \phi(x) \ . \end{aligned} \tag{2.5.1}$$

Then $\frac{\partial u}{\partial t} \leq 0$ for all $x \in D$, $t > 0$. Similarly if u satisfies the boundary-initial value problem (2.5.1) and $u(x,0) = \psi(x)$, where ψ is a lower solution, then $u_t \geq 0$ for $x \in D$ and $t \geq 0$.

Before proving Theorem 2.5.1 let us note the following: We construct a sequence of iterations $\{u_n\}$ by $u_0(x,t) = \phi(x)$,

$$Lu_n - \Omega u_n - \frac{\partial u_n}{\partial t} = -[f(x,u_{n-1}) + \Omega u_{n-1}]$$

$$u_n\Big|_{\partial D} = g \qquad\qquad u_n(x,0) = \phi(x) \ .$$

If $L\phi + f(x,\phi) \leq 0$ then $\phi(x) \geq u_1(x,t) \geq \ldots \geq u(x,t)$, where u is the solution of the boundary initial value problem (2.5.1). In the limit, $u(x,t) = \lim_n u_n(x,t)$, so

$$u(x,t) \leq \phi(x) \tag{2.5.2}$$

if ϕ is an upper solution. Now we can begin the proof:

Proof: Put

$$w_h(x,t) = \frac{u(x,t+h) - u(x,t)}{h}$$

where $h > 0$ and $u(x,t)$ is the solution of (2.5.1) .

We easily see that

$$Lw_h + \frac{f(x,u(x,t+h)) - f(x,u(x,t))}{h} - \frac{\partial w_h}{\partial t} = 0$$

Now by the mean value theorem,

$$f(x,b) - f(x,a) = \int_0^1 f'_u(x,\tau b + (1-\tau)a)d\tau(b-a) \quad .$$

Letting $b = u(x,t+h)$ and $a = u(x,t)$ we get

$$\frac{f(x,u(x,t+h)) - f(x,u(x,t))}{h} = \xi(x,t,h)w_n(x,t)$$

where

$$\xi(x,t,h) = \int_0^1 f'_u(x,\tau u(x,t+h) + (1-\tau)u(x,t))d\tau \quad .$$

Therefore $w_h(x,t)$ satisfies

$$Lw_h + \xi\, w_h - \frac{\partial w_h}{\partial t} = 0$$

and also

$$w_h(x,t)|_{\partial D} = \frac{1}{h}\{u(x,t+h)|_{\partial D} - u(x,t)|_{\partial D}\} = 0$$

$$w_h(x,0) = \frac{u(x,h) - u(x,0)}{h} \leq 0$$

by (2.5.2). Now, however, the maximum principle applied to w_h implies that $w_h(x,t) \leq 0$ for $t \geq 0$. Therefore

$$\frac{\partial u}{\partial t} = \lim_{h \to 0^+} w_h(x,t) \leq 0 \quad . \qquad \text{Q.E.D.}$$

For the next steps we need the comparison theorem for parabolic equations:

Theorem 2.5.2: Let D be a bounded domain in R^n and let $E = D \times (0,T]$. Suppose u is a solution of

$$Lu + f(x,u) - \frac{\partial u}{\partial t} = 0$$

in E with boundary and initial conditions

$$u(x,0) = u_0(x)$$

$$u(x,t)\big|_{\partial D} = g(x) \quad .$$

Let z and Z satisfy the inequalities

$$LZ + f(x,Z) - \frac{\partial Z}{\partial t} \leq 0 \qquad Z(x,0) \geq u_0(x), \; Z(x,t)\big|_{\partial D} \geq g(x)$$

$$Lz + f(x,z) - \frac{\partial z}{\partial t} \geq 0 \qquad z(x,0) \leq u_0(x), \; z(x,t)\big|_{\partial D} \leq g(x)$$

Then $z(x,t) \leq u(x,t) < Z(x,t)$ in E.

This theorem is a special case of Theorem 12, p. 187 in Protter and Weinberger's book.

2.6 The Initial Value Problem. (cont.)

Suppose that ϕ is an upper solution and ψ a lower solution with $\psi < \phi$. We let u satisfy

$$Lu + f(x,u) - u_t = 0$$
$$u\big|_{\partial D} = 0 \qquad u(x,0) = u_0(x) \tag{2.6.1}$$

where $\psi(x) \leq u_0(x) \leq \phi(x)$. The functions $\phi(x,t) \equiv \phi(x)$ and $\psi(x,t) \equiv \psi(x)$ satisfy

$$L\phi + f(x,\phi) - \frac{\partial\phi}{\partial t} \leq 0 \qquad \phi(x)\Big|_{\partial D} \geq 0$$

$$L\psi + f(x,\psi) - \frac{\partial\psi}{\partial t} \geq 0 \qquad \psi(x)\Big|_{\partial D} \leq 0$$

so by the comparison theorem, $\psi(x) \leq u(x,t) \leq \phi(x)$ for all $t > 0$.

In particular, if $u(x,0) = \phi(x)$ then

$$u_t \leq 0 \qquad \text{and} \qquad \psi(x) \leq u(x,t) \leq \phi(x) \ . \tag{2.6.2}$$

We shall prove the following

<u>Theorem 2.6.1</u>: <u>Under the above assumptions, $\lim_{t \to \infty} u(x,t) = \hat{u}(x)$ exists and is a solution of the Dirichlet problem</u>

$$L\hat{u} + f(x,\hat{u}) = 0 \qquad \hat{u}\Big|_{\partial D} = g(x) \ .$$

<u>Proof</u>: By the monotone iteration procedure we can construct, for any $T > 0$, a regular solution to the boundary-initial value problem (2.6.1) for $0 \leq t \leq T$, so the solution exists for all $t > 0$ and satisfies (2.6.2). Since $u(x,t)$ is monotone non-increasing, $\hat{u}(x) = \lim_{t \to \infty} u(x,t)$ exists, and $\psi(x) \leq \hat{u}(x) \leq \phi(x)$.

Let L^* be the adjoint operator and take the inner product of (2.6.1) with a smooth function $\xi(x)$ with $\xi = 0$ on ∂D :

$$(Lu,\xi) + (f(x,u),\xi) - (\xi,u_t) = 0 \ ,$$

$$(u,L^*\xi) + (f(x,u),\xi) - (\xi,u_t) = 0 \ .$$

So

$$\frac{1}{T}\int_0^T (u(x,t),L^*\xi)dt + \frac{1}{T}\int_0^T (f(x,u(x,t)),\xi)$$

$$- \frac{1}{T}\int_0^T (\xi,u_t)dt = 0 \quad ,$$

Now let $T \to \infty$. Since $u(x,t) \to \hat{u}(x)$, $\lim_{T \to \infty} \frac{1}{T}\int_0^T u(x,t)dt = \hat{u}(x)$,

and

$$\lim_{T \to \infty} \frac{1}{T}\int_0^T (u(x,t),L^*\xi)dt$$

$$= \lim_{T \to \infty} \left(\frac{1}{T}\int_0^T u(x,t)dt,\ L^*\xi\right) = (\hat{u}(x),L^*\xi)$$

by the Lebesgue dominated convergence theorem.
(u is bounded, so $\frac{1}{T}\int_0^T u(x,t)dt$ is bounded, and $\lim_{T \to \infty} \frac{1}{T}\int_0^T u(x,t)dt = \hat{u}(x)$) .

Similarly,

$$\lim_{T \to \infty} \frac{1}{T}\int_0^T (\xi,u_t)dt = \lim_{T \to \infty} \frac{1}{T}\int_0^T \frac{\partial}{\partial t}(\xi,u)dt$$

$$= \lim_{T \to \infty} \frac{(\xi,u(T)) - (\xi,u(0))}{T} = 0$$

So in the limit we get

$$(\hat{u},L^*\xi) + (f(x,\hat{u}),\xi) = 0 \qquad (2.6.3)$$

for every smooth function ξ which vanishes on ∂D .

Now we have to show that $\hat{u}$ is a regular solution of the Dirichlet problem. First we note that L and L^* are invertible. This is a consequence of the Fredholm alternative, Theorem (2.2.4); for, by the maximum principle, the boundary value problems $Lw = 0$, $w = 0$ on ∂D and $L^*w = o$, $w = 0$ on ∂D have only the trivial solution. Let G be the inverse of L. By a standard result of functional analysis (see Riesz - Nagy, p. 304) G^* is the inverse of L^*. Putting $w = -Gf(x,\hat{u})$ we have

$$(w,L^*\xi) = -(Gf,L^*\xi) = -(f(x,\hat{u}),\ G^*L^*\xi)$$

$$= -(f(x,\hat{u}),\xi) .$$

But from (2.6.3) we get

$$(\hat{u},L^*\xi) = -(f(x\hat{u}),\xi) = (w,L^*\xi) .$$

Therefore

$$(\hat{u} - w,L^*\xi) = 0$$

for all smooth ξ which vanish on ∂D. In particular this holds for test functions $\xi = G^*\eta$ where η is smooth and bounded. Hence $(\hat{u} - w,L^*G^*\eta) = (\hat{u} - w,\eta) = 0$ for all smooth η, and $\hat{u} = w$ a.e. But w is continuous, so modifying $\hat{u}$ on a set of measure zero if necessary we get

$$\hat{u}(x) = w(x) = -Gf(x,\hat{u}) .$$

From Theorem (2.2.3) we can conclude that $\hat{u}$ is in the class $C_{2+\alpha}(\bar{D})$,

$\hat{u} = 0$ on ∂D , and $L\hat{u} + f(x,\hat{u}) = 0$ in D .

We now prove:

Theorem 2.6.2: Let $\hat{u}$ be a solution of

$$L\hat{u} + f(x,\hat{u}) = 0 \qquad u\big|_{\partial D} = g \tag{2.6.4}$$

Let ϕ and ψ be upper and lower solutions of (2.6.4) with $\psi \le \hat{u} \le \phi$ and let $\hat{u} = \lim_n T^n\phi = \lim_n T^n\psi$. Then $\hat{u}$ is asymptotically stable and all solutions of (2.6.1) with initial data $\psi \le u_0(x) \le \phi(x)$ tend to $u(x)$ as $t \to \infty$.

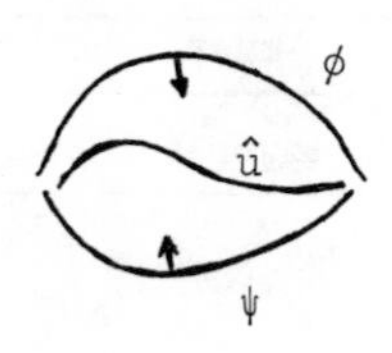

Proof: Let $\phi(x,t)$, $\psi(x,t)$, and $u(x,t)$ denote solutions of the initial value problem with initial data $\phi(x)$, $\psi(x)$, and $u_0(x)$ respectively. By the comparison theorem $\psi(x,t) \le u(x,t) \le \phi(x,t)$. By Theorem 2.6.1 , $\lim_{t \to \infty} \phi(x,t) = \hat{\phi}(x)$ and $\lim_{t \to \infty} \psi(x,t) = \hat{\psi}(x)$ where $\hat{\phi}$ and $\hat{\psi}$ are solutions of the stationary problem. Since $\lim_n T^n\phi = \lim_n T^n\psi$ there is only one solution between $\phi(x)$ and $\psi(x)$, and that is $\hat{u}(x)$. So $\hat{\phi} = \hat{\psi} = \hat{u}$ and the solution $u(x,t)$ is squeezed into $\hat{u}$ as t tends to infinity.

2.7 Serrin's Sweeping Principle.

Not only is it nice to know where solutions are, it is also a good thing to know where they are not. The following theorem of Serrin [1] (Theorem 2, p. 15), provides a technique for doing just that.

<u>Theorem 2.7.1 (Serrin): Suppose $v_\lambda = v(x,\lambda)$ is a family of upper solutions which is increasing in λ , $a \le \lambda \le b$:</u>

$$Lv_\lambda + f(x,v_\lambda) \le 0$$

<u>If u is any solution to</u>

$$Lu + f(x,u) = 0 \tag{2.7.1}$$

<u>such that $u \le v_b$ in D and $u \le v_a$ on ∂D then we have either $u \equiv v_a$ or $u < v_a$ in D . In particular, if $v_a\big|_{\partial D} \ge g$ then any solution of (2.7.1) satisfying $u|_{\partial D} = g$ and $u \le v_b$ must also satisfy $u < v_a$. Similarly, if v_λ is a family of lower solutions</u>

$$Lv_\lambda + f(x,v_\lambda) \ge 0$$

<u>$a \le \lambda \le b$ then any solution of $Lu + f(x,u) = 0$ for which $u \ge v_a$ and $u|_{\partial D} \ge v_b$ must satisfy $u > v_b$ in D .</u> Serrin's theorem actually applies to general second order nonlinear elliptic operators.

Let us return to our simple nonlinear problem

$$(\Delta + \mu)u - u^3 = 0 \qquad u\big|_{\partial D} = 0 \ . \tag{2.7.2}$$

We already know that for $\mu > \lambda_1$ (2.7.2) has two non-trivial solutions w and $-w$, with $w > 0$. Let us show that w is the only positive solution of (2.7.2)

Put $v_\lambda(x) = \lambda w$. We have

$$(\Delta + \mu)v_\lambda - v_\lambda{}^3 = \lambda\ (1 - \lambda^2)w^3 \quad \begin{matrix} > 0 & \text{if } \lambda < 1 \\ < 0 & \text{if } \lambda > 1 \end{matrix} \ .$$

So for $\lambda > 1$ the v_λ are upper solutions while for $\lambda < 1$ the v_λ are lower solutions.

Now let z be any other smooth non-negative solution of (2.7.2). We claim that $z > 0$ in the interior of D. In fact, since $\mu > 0$, in a neighborhood of $z = 0$ we have

$$\Delta z = -\mu z + z^3 \leq 0$$

for $z \geq 0$ and small. But by the maximum principle z could not then have a local interior minimum unless z were identically constant. Similarly, since $z = 0$ on ∂D we have $\Delta z \leq 0$ in a neighborhood of the boundary. Therefore, by the maximum principle at the boundary, $\frac{\partial z}{\partial \nu} < 0$ on ∂D. Since ∂D is compact, $\frac{\partial z}{\partial \nu} \leq -\delta > 0$ on ∂D for some $\delta > 0$.

Similarly $\frac{\partial w}{\partial \nu} \leq -\delta < 0$ on ∂D, where w is the given positive solution. There are two cases to consider: $\max z > \max w$ and $\max z < \max w$. In the first case there exists λ sufficiently large so that $z < \lambda w$ everywhere in $\bar{D}$. (Note that since $\frac{\partial w}{\partial \nu} \leq -\delta < 0$, $\lambda \frac{\partial w}{\partial \nu} < \frac{\partial z}{\partial \nu}$ on ∂D for sufficiently large λ, so z/w is bounded on $\bar{D}$.) We now apply Serrin's theorem to the family $v_\lambda = \lambda w$. For $\lambda = 1$, $z \leq v_1 = w$ on ∂D while $z < v_\lambda$ for sufficiently large λ. So either $z < v_1 = w$ or $z = w$, both of which are contradictions. By similar arguments, applying the second statement of Serrin's theorem, we can eliminate the second possibility.

<u>Exercise</u>: Show that $\Delta u + u^2 = 0$, $u\big|_{\partial D} = 0$ has at most one nontrivial solution, which must be positive and unstable. Discuss the behavior of solutions of the initial value problem

$$\Delta u + u^2 = u_t$$

$$u\big|_{\partial D} = 0 \qquad u(x,0) = u_0(x) .$$

2.8 A Singular Perturbation Problem.

The following singular perturbation problem leads to another way to construct upper and lower solutions:

$$\beta u'' - u' + f(u) = 0 \qquad 0 \leq x \leq 1$$

$$-u'(0) + au(0) = 0 \qquad (2.8.1)$$

$$u'(1) = 0$$

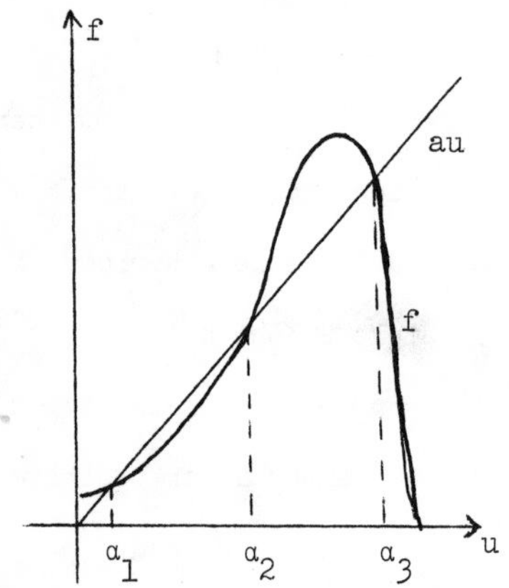

Fig. 2.8.1

This problem was treated by D. S. Cohen [1] in connection with problems arising in chemical reactor theory. The function f has the qualitative behavior shown at the left. β is supposed to be a small parameter. According to singular perturbation theory the solution of (2.8.1) has, for small β, a second derivative of order 1 in the interior of $0 < x < 1$. So in the interior the term $\beta u''$ can be neglected: The solution in the interior is given approximately by

$$u' = f(u) \qquad (2.8.2)$$

There are possible boundary layers — that is, regions where u'' is not small — at the boundaries $x = 0$ and $x = 1$. It turns out in this case, however, that no boundary layer occurs at $x = 0$. We choose $u(0) = \alpha$ so that (2.8.2) and the boundary condition at $x = 0$ in (2.8.1) is satisfied: Taking $u(0) = \alpha$ we get $u'(0) = f(\alpha)$ from (2.8.2) and $u'(0) = a\alpha$ from (2.8.2), hence

$$a\alpha = f(\alpha)$$

The solutions of this equation are indicated graphically in Fig. 2.8.1. For appropriate choices of a we get three solutions α_1, α_2, and α_3. Let us show that, for small β, the roots α_1 and α_3 lead to stable solutions of (2.8.1).

We solve the first order problem

$$\begin{aligned} - v' + f(v) &= -\varepsilon \\ v(0) &= \alpha + \delta \end{aligned} \tag{2.8.3}$$

where ε and δ are parameters and $\alpha = \alpha_1$ or α_3. Denoting by L the operator $Lu = \beta u'' - u'$ we have

$$\begin{aligned} Lv + f(v) &= \beta v'' - v' + f(v) \\ &= \beta v'' - \varepsilon , \end{aligned}$$

$$\begin{aligned} - v'(0) + av(0) &= - \varepsilon - f(v(0)) + av(0) \\ &= - \varepsilon +[a(\alpha + \delta) - f(\alpha + \delta)] \\ &= - \varepsilon + a\alpha - f(\alpha) + [a\delta - f'(\alpha)\delta] + O(\delta^2) \\ &= - \varepsilon + \delta[a - f'(\alpha)] + O(\delta^2) . \end{aligned}$$

Now for $\alpha = \alpha_1$ or α_3 we have $a - f'(\alpha) > 0$ (see Fig. 2.8.1), so taking ϵ sufficiently small, say

$$\epsilon = \frac{\delta}{2}(a - f'(\alpha)) \quad ,$$

we have (denoting the solution of (2.8.3) by v_δ)

$$\begin{aligned} Lv_\delta + f(v_\delta) &< 0 \\ - v_\delta'(0) + av_\delta(0) &> 0 \end{aligned} \tag{2.8.4}$$

for $\delta > 0$ and β sufficiently small. If $\delta < 0$ these inequalities are reversed. (Note that v_δ'' remains bounded for small $|\delta|$, so $\beta v_\delta'' = O(\beta)$. We first pick $\delta_0 > 0$ and then restrict β to be so small that 2.8.4 holds for $|\delta| < \delta_0$.)

For small $\delta > 0$ v_δ is to be an upper solution and $v_{-\delta}$ a lower solution. We claim that $v_{-\delta} < v_\delta$ on $0 \leq x \leq 1$. In fact, $v_\delta(0) - v_{-\delta}(0) = 2\delta > 0$. Let x_0 be the first value of x at which $v_\delta(x_0) = v_{-\delta}(x_0)$. We have $v'_{-\delta}(x_0) \leq v_\delta'(x_0)$, hence, from (2.8.3) $f(v_\delta) + \epsilon \leq f(v_{-\delta}) - \epsilon$, which implies that $\epsilon \leq -\epsilon$, a contradiction. So $v_\delta(x) > v_{-\delta}(x)$ on [0,1].

We have one final matter to deal with, and that is the boundary condition at $x = 1$. For $\delta > 0$ we are all right, since $v_\delta'(1) = f(v_\delta(1) + \epsilon > 0$; so for $\delta > 0$, v_δ really is an upper solution. To actually get the lower solution we take

$$u_{-\delta}(x) = v_{-\delta}(x) - \beta v'_{-\delta}(1)e^{-(1-x)/\beta} \quad .$$

Then $u'_{-\delta}(1) = 0$ and $u_{-\delta}(x) \leq v_{-\delta}(x)$. This gives us the requisite lower solution.

For $\alpha = \alpha_1$ or α_3 we see that it is possible, for small β, to construct a lower solution below and an upper solution above the singular perturbation solution. Therefore, for small β, there is a stable solution in the vicinity of the singular perturbation solution

$$v(x) = v_0(x) - \beta v_0'(1) e^{-(1-x)/\beta} .$$

For $\alpha = \alpha_2$ the equation and boundary conditions work against each other; this time we get a lower solution above and an upper solution below the singular perturbation solution — thus indicating, formally at least, that the middle solution might be unstable.

Exercises:

1. Prove the principle of linearized stability for problems of the form

$$Lu + f(x,u) - u_t = 0$$

   + initial, boundary conditions.

   That is, if $\hat{u}$ is a stationary solution, $L\hat{u} + f(x,\hat{u}) = 0$, then $\hat{u}$ is stable if all eigenvalues of the operator

$$L + f_u'(x,\hat{u})$$

   have positive real parts, and unstable if some eigenvalues have negative real parts. (Hint: consider $\hat{u} + \epsilon \varphi_1$, where φ_1 is the principal eigenfunction of $L + f_u'(x,\hat{u})$)

2. Carry through the necessary modifications for problem (2.8.1) with the boundary conditions at $x = 1$ replaced by $u(1) = 0$.

3. Consider the analogue of problem (2.8.1) on a cylindrical domain $0 \leq z \leq L$, $0 \leq r \leq 1$. Construct the singular perturbation solutions and justify their validity in the case of stable solutions. (viz. $\alpha = \alpha_1$ and α_3)

Notes

Monotone iteration schemes such as discussed here were developed in a series of papers by Cohen, H. B. Keller, and their associates. See Cohen and H. B. Keller, Cohen and Simpson, H. B. Keller [1,2,3]. Theorem 2.3.1 was first proved by Amann [1]. The singular perturbation argument was given by Cohen [1]; see also H. B. Keller [1,2]. Monotone methods have been used in connection with other problems, such as the neutron transport equation Pazy and Rabinowitz. For the development of stability theory using upper and lower solutions, see Sattinger [4], Lee and Luss [1]. Bazley and Zwahlen, and Reeken [1], have considered a class of problems arising from the Hartree equation for the Helium atom.

III

Functional Analysis

In order to go further into the subject of bifurcation and stability theory it is first necessary to present some basic facts of functional analysis. The basic concepts and tools needed are: 1° the Riesz-Schauder theory of compact operators, 2° the Frechet derivative of an operator, 3° the implicit function theorem in a Banach space.

We assume the reader is familiar with the basic definitions and examples of Hilbert and Banach spaces, subspaces, projections, bounded linear operators on Banach spaces, and their adjoints. Two Banach spaces which are important in nonlinear partial differential equations are the spaces $W_{k,p}$ and $C_{k+\alpha}$, introduced in § 2.2.
A linear functional f^* on a Banach space $\mathfrak{B}$ is a mapping from $\mathfrak{B}$ to the complex numbers such that

(i) $|f^*(x)| \leq \|x\|$

(ii) $f^*(\alpha x + \beta y) = \alpha f^*(x) + \beta f^*(y)$

for all complex numbers α and β and all vectors x,y in $\mathfrak{B}$. The set of all continuous linear functionals on $\mathfrak{B}$ forms another Banach space $\mathfrak{B}^*$ with norm given by

$$\|f^*\| = \sup_{\|x\|=1} |f^*(x)| \quad .$$

It is often convenient to use the notation $f^*(x) = \langle x,f^*\rangle$.

More generally, the class of linear transformations from $\mathfrak{B}_1$

to $\mathcal{B}_2$ is a Banach space under the norm

$$\|L\| = \sup_{\|x\|_1=1} \|Lx\|_2 \quad .$$

Here $\| \ \|_i$ denotes the norm in the Banach space $\mathcal{B}_i$.
If L is a bounded linear transformation from a Banach space $\mathcal{B}$ to itself, we denote the adjoint, which maps $\mathcal{B}^*$ to $\mathcal{B}^*$ and satisfies $\langle Lx,f^*\rangle = \langle x,L^*f^*\rangle$, by L^* .

3.1 The Riesz-Schauder Theory.

Definition: Let T be a continuous (not necessarily linear) operator on a Banach space $\mathcal{B}$. T is said to be completely continuous if for every bounded set ω in $\mathcal{B}$, the set $T(\bar{\omega})$ is compact.

For linear completely continuous (sometimes called compact) operators on a Banach space we have the Riesz-Schauder theory, the essentials of which follow below: (T is a compact linear operator)

Fredholm Alternative: Either the homogeneous problem

$$(I - T)\varphi = 0 \tag{3.1.1}$$

has a non-trivial solution, or the equation

$$(I - T)u = f \tag{3.1.2}$$

is boundedly invertible. That is, for any f (3.1.2) has a unique solution u and $\|u\| \leq C\|f\|$, where the constant C is independent of f .

If the homogeneous equation has k non-trivial solutions then so does the adjoint equation

$$(I - T^*)\varphi^* = 0$$

and (3.1.2) is solvable if and only if $\langle f,\varphi^*\rangle = 0$ for all solutions φ^* of the homogeneous adjoint problem.

Null and Range Spaces:

Define

$$\mathfrak{M}_n = \{g : (I - T)^n g = 0\} ,$$

$$\mathfrak{N}_n = \{f : f = (I - T)^n \varphi \quad \text{for some} \quad \varphi \in \mathfrak{B}\}$$

Then $\{0\} = \mathfrak{M}_0 \subset \mathfrak{M}_1 \subset \mathfrak{M}_2 \subset \dots$ and $\mathfrak{B} = \mathfrak{N}_0 \supset \mathfrak{N}_1 \supset \dots$. If T is compact there is an integer ν such that

$$\mathfrak{M}_{\nu-1} \subset \mathfrak{M}_\nu \qquad \mathfrak{N}_{\nu-1} \supset \mathfrak{N}_\nu \qquad \text{(strictly)} ,$$

and

$$\mathfrak{M}_\nu = \mathfrak{M}_{\nu+1} = \dots \qquad \mathfrak{N}_\nu = \mathfrak{N}_{\nu+1} = \dots$$

The dimension of $\mathfrak{M}_\nu$ is finite and

(i) $\mathfrak{B} = \mathfrak{M}_\nu + \mathfrak{N}_\nu$, $\mathfrak{M}_\nu \cap \mathfrak{N}_\nu = \phi$

(ii) T leaves $\mathfrak{M}_\nu$ and $\mathfrak{N}_\nu$ invariant.

The first statement means that any $f \in \mathfrak{B}$ has a unique decomposition $f = u + v$ where $u \in \mathfrak{M}_\nu$ and $v \in \mathfrak{N}_\nu$.

Projections: Define projections P and Q $(Q = I - P)$ by $Pf = u$, $Qf = v$, where $f = u + v$, $u \in \mathfrak{M}_\nu$, $v \in \mathfrak{N}_\nu$. P and Q commute with T and furthermore, $(I - T)Q$ has a bounded inverse from N_ν to itself. In fact, we may regard $\mathfrak{N}_\nu$ as a Banach space on which T is compact; since $(I - T)$ maps $\mathfrak{N}_\nu$ into itself, and since $I - T$ has no nontrivial solutions on $\mathfrak{n}_\nu$ we may apply the Fredholm alternative to $(I - T)$ on $\mathfrak{n}_\nu$. Thus there is a bounded linear transformation S such that

$$S(I - T) = I \quad \text{on} \quad \mathfrak{n}_\nu ,$$

or

$$S(I - T)Q = Q .$$

Since T leaves the finite dimensional subspace $\mathfrak{m}_\nu$ invariant we can construct a Jordan canonical basis for $\mathfrak{m}_\nu$ with respect to T. Thus we have

$$T\varphi_1 = \varphi_1$$

$$T\varphi_2 = \varphi_2 + \varphi_1$$

$$T\varphi_3 = \varphi_3 + \varphi_2 , \dots$$

and possibly other such chains of eigenfunctions.

Since P is a projection onto the finite dimensional subspace $\mathfrak{m}_\nu$ we can write

$$Pu = \sum_{i=1}^{d} C_i(u)\varphi_i ,$$

where $d = \dim \mathfrak{m}_\nu$. Since the φ_i are linearly independent and P is linear we must have $C_i(\alpha u + \beta v) = \alpha C_i(u) + \beta C_i(v)$. Therefore the C_i are linear functionals on $\mathfrak{B}$ and we may write $C_i(u) = \langle u, \varphi_i^* \rangle$, for some φ_i^* in $\mathfrak{B}^*$.

Since P is a projection onto $\mathfrak{M}_\nu$ and $\varphi_i \in \mathfrak{M}_\nu$, $P\varphi_i = \varphi_i$. This immediately leads to

$$\langle \varphi_i, \varphi_j^* \rangle = \delta_{ij} \quad .$$

Suppose $\nu = 1$ and $T\varphi = \varphi$. Then

$$Pu = \langle u, \varphi^* \rangle \varphi ,$$

and from $TPu = PTu$ and $T\varphi = \varphi$ we conclude that

$$\langle u, \varphi^* \rangle \varphi = \langle u, \varphi^* \rangle T\varphi = \langle Tu, \varphi^* \rangle \varphi = \langle u, T^* \varphi \rangle \varphi$$

for all u . Hence

$$\langle u, \varphi^* - T^* \varphi \rangle = 0$$

for all u , and

$$\varphi^* - T^* \varphi = 0.$$

More generally, it is a simple algebraic matter to show that the elements φ_i^* form a Jordan basis for T^* . See exeecise 1 .

A complex number λ is called a characteristic value of T if $(I - \lambda T)\varphi = 0$ has non-trivial solutions. An eigenvalue of T is a complex number μ such that $T\varphi = \mu\varphi$ for some nontrivial φ . The characteristic values are the reciprocals of the eigenvalues.

<u>Solvability</u>: Consider the functional equation

$$(I - T)u = f \tag{3.1.3}$$

in the case where $(I - T)\varphi = 0$ has non-trivial solutions.

Define $\mathfrak{n}^*_{\nu^*}$ for $(I - T^*)$ in the same way that $\mathfrak{n}_\nu$ was defined for the operator $I - T$. Then $\nu^* = \nu$ and $\dim \mathfrak{n}^*_{\nu^*} = \dim \mathfrak{n}_\nu$.
For the solvability of (3.1.3) it is sufficient that $\langle f,\varphi^*\rangle = 0$ for all $\varphi^* \in \mathfrak{n}_1^*$, even if $\nu > 1$. For example, if $\nu = 2$ and $\dim \mathfrak{n}_2 = 2$ then (Exercise 1)

$$(I - T)\varphi_1 = 0 \qquad\qquad (I - T)\varphi_2^* = 0$$

$$(I - T)\varphi_2 = \varphi_1 \qquad\qquad (I - T)\varphi_1^* = \varphi_2^*$$

To solve (3.1.3) where $\langle f,\varphi_2^*\rangle = 0$ write $f = Pf + Qf$ where $Pf \in \mathfrak{m}_2$ and $Qf \in \mathfrak{n}_2$. There is a unique u_2 in $\mathfrak{n}_2$ such that

$$(I - T)^2 u_2 = Qf$$

The general solution of (3.1.3) is then

$$u = (I - T)u_2 + \langle f,\varphi_1^*\rangle\varphi_2 + a\varphi_1$$

where a is arbitrary. (The reader should check this)

<u>Exercises</u>:

1. Suppose $\nu = 2$ and $T\varphi_1 = \varphi_1$, $T\varphi_2 = \varphi_2 + \varphi_1$. Show that φ_1^* and φ_2^* form a Jordan basis for T^*. Do the problem for arbitrary ν.

2. If λ is a characteristic value for T then $\bar{\lambda}$ is a characteristic value of T^*. Let T have simple $(\nu = 1)$ characteristic values $\pm i$. Show there is a basis φ_1, φ_2 for the subspace corresponding to $\pm i$ such that

$$T\varphi_2 = \varphi_1 \qquad\qquad T\varphi_1 = -\varphi_2$$

What happens to the dual functions φ_1^* and φ_2^* under T^*?

*3. Show that $H_\beta \subset H_\alpha$ if $\alpha > \beta$. Show that bounded sets in H_α are precompact in H_β. That is, if $\|f_n\|_\alpha \leq M$ then there exists a subsequence $f_{n'}$ such that $\|f_{n'} - f_{m'}\|_\beta \underset{n',m'\to\infty}{\to} 0$.

4. Prove $\|fg\|_\alpha \leq \|f\|_\alpha \cdot \|g\|_\alpha$.

*5. Prove H_α is a Banach space. (Hint: apply the Arzela-Ascoli theorem)

6. Let $\beta = (\beta_1, \ldots \beta_n)$ and $D^\beta = \dfrac{\partial|\beta|}{\partial x_1^{\beta_1} \ldots \partial x_n^{\beta_n}}$, where $|\beta| = \beta_1 + \ldots + \beta_n$. Define $C_{k+\alpha}^{(\Omega)}$ to be the set of all functions u on Ω for which $\|u\|_{k+\alpha} < +\infty$, where

$$\|u\|_{k+\alpha} = \sum_{|\beta|\leq k} \|D^\beta u\|_\alpha$$

Show $C_{k+\alpha}$ is a Banach space, that every bounded set in $C_{k+\alpha}$ contains a subsequence which converges in $C_{k+\alpha'}$ for any $\alpha' < \alpha$, and that $\|fg\|_{k+\alpha} \leq \|f\|_{k+\alpha}\|g\|_{k+\alpha}$.

7. Prove that the class of bounded linear operators from Banach spaces $\mathcal{B}_1$ to $\mathcal{B}_2$, denoted by $\mathcal{L}(\mathcal{B}_1, \mathcal{B}_2)$, is itself a Banach space with the norm

$$\|A\| = \sup_{\|x\|_1=1} \|Ax\|_2 .$$

8. Let $C_0(\Omega)$ be the space of continuous functions on Ω with norm $\|u\|_0 = \sup_{x\in\Omega} |u(x)|$. Show C_0 is a Banach space.

3.2 <u>The Frechet Derivative</u>:

Let $\mathcal{B}_1$ and $\mathcal{B}_2$ be Banach spaces (possibly the same) and let T be a continuous mapping from $\mathcal{B}_1$ to $\mathcal{B}_2$.

<u>Definition</u>: <u>We say that T has a Frechet derivative $A(x_0)$ at a point x_0 in $\mathcal{B}_1$ if $A(x_0)$ is a linear operator such that the operator R,</u>

$$R(x_0,v) = T(x_0 + v) - T(x_0) - A(x_0)v \quad ,$$

<u>satisfies $\|R(v)\|_2 = o(\|v\|_1)$ as $\|v\|_1 \to 0$. Here $\|\ \|_1$ and $\|\ \|_2$ denote the norms in $\mathcal{B}_1$ and $\mathcal{B}_2$ respectively. The notation $\|R(v)\|_2 = o(\|v\|_1)$ means that</u>

$$\lim_{\|v\|_1 \to \infty} \frac{\|R(v)\|_2}{\|v\|_1} = 0 \ .$$

<u>We say T is continuously differentiable on an open set $\Omega \subset \mathcal{B}_1$ if $A(x_0)$ is a continuous mapping from Ω to the Banach space $\mathcal{L}(\mathcal{B}_1,\mathcal{B}_2)$ and if R is continuous and satisfies $\|R(x_0,v)\|_2 = o(\|v\|_1)$ uniformly as x_0 ranges over closed subsets of Ω.</u>

Note that the Frechet derivative of any linear operator T is T itself; R, the remainder operator, is then identically zero.

<u>Examples</u>:

1. $T : C_0 \to C_0$ is defined by $T(u) = u^2$. Then

$$T(u_0 + v) = u_0^2 + 2u_0v + v^2$$

so $T'(u_0)$ is the linear operator $T'(u_0)v = 2u_0v$, and $R(u_0,v) = v^2$. T is continuously differentiable everywhere in C_0.

2. $T : C_1 \to C_0$ is defined by $T(u) = uu_x$. Then

$$T(u_0 + v) = u_0 u_{0x} + u_0 v_x + v u_{0x} + v v_x$$

So $T'(u_0)v = u_0 v_x + v u_{0x}$ and $R(u_0, v) = v v_x$. T is continuously Frechet differentiable everywhere on C_1 .

3. Let $B = B(u,v)$ be a bilinear operator from $\mathcal{B}_1 \times \mathcal{B}_1$ to $\mathcal{B}_2$:

$$B(\alpha u + \beta v, w) = \alpha B(u,w) + \beta B(v,w)$$

and similarly for the right hand argument.

We also assume

$$\|B(u,v)\|_2 \leq M\|u\|_1\|v\|_1 \quad .$$

Let $T(u) = B(u,u)$. Then T is Frechet differentiable with $T'(u_0)v = B(u_0,v) + B(v,u_0)$.

4. Let F be a differentiable mapping from $R^n \to R^m$:

$$F(x) = (f_1(x_1, \ldots, x_n) \;, \ldots \; f_m(x_1, \ldots, x_n)) \; .$$

The Frechet derivative of F at $u = (u_1, \ldots, u_n)$ is a linear operator

$$F'(u)v = \sum_{j=1}^{n} \frac{\partial f_i}{\partial x_j} v_j$$

where the $\frac{\partial f_i}{\partial x_j}$ are evaluated at $(u_1, \ldots, u_n)$.

Taylor Series

In a number of situations a Taylor series expansion exists for a given nonlinear operator. Let B_k be a multilinear operator of degree k : that is

$$B_k(u_1, \dots u_k)$$

is a continuous operator from $\underbrace{\mathfrak{B}_1 \times \dots \times \mathfrak{B}_1}_{k \text{ times}}$ to $\mathfrak{B}_2$ which is linear in each argument, and there is a constant M such that

$$\|\mathfrak{B}_k(u_1, \dots u_k)\|_2 \leq M\|u_1\|_1 \dots \|u_n\|_1 \ .$$

<u>Definition</u>: <u>We say that an operator $F : \mathfrak{B}_1 \to \mathfrak{B}_2$ is k times continuously differentiable in a region Ω if the following holds</u>: <u>There exist j-linear operators</u> $(j = 1, \dots k)$ $F_1(x_0;v)$, $F_2(x_0;v,v)$, ... $F_k(x_0;v, \dots v)$ <u>which are continuous from</u> $\Omega \times \mathfrak{B}_1 \times \dots \times \mathfrak{B}_1$ to $\mathfrak{B}_2$ <u>such that the operator</u>

j

$$R(x_0,v) = F(x_0 + v) - F(x_0) - F_1(x_0,v) \dots - F_k(x_0;v, \dots v)$$

<u>satisfies</u> $\|R(x_0,v)\|_2 = o(\|v\|_1^k)$ <u>uniformly on closed subsets of Ω</u>.

An operator F is said to be analytic at a point x_0 if F has the representation

$$F(x_0 + v) = F(x_0) + \sum_{j=1}^{\infty} F_j(v)$$

with a convergent series for $\|v\|_1 < \delta$; here F_j is a j-linear operator.

If $\|F_j(v)\|_2 \leq M_j\|v\|_1^j$ it is enough to assume that the series $\sum_{j=1}^{\infty} M_j\delta^j$ has a positive radius of convergence in order to guarantee that F is analytic in $\|v\| < \delta$.

The "Derived operator". Sometimes it is necessary to consider the case where F is a continuous mapping from a Banach space $\mathcal{B}_1$ to $\mathcal{B}_2$, but is not continuous from $\mathcal{B}_1$ to itself. The operator $F(u) = uu_x$ furnishes such an example. F has a Frechet derivative when considered as a mapping from, say, C^1 to C^0 , but not when considered as a mapping from C^0 to itself. The Frechet derivative (from C^1 to C^0) would be

$$F'(u_0)v = u_0 v_x + vu_{0,x}$$

Thus, considered as a mapping from C^0 to itself, $F'(u_0)$ is an unbounded linear operator. We shall refer to the operator $F'(u_0)$ in this case as the derived operator. Algebraically it has the same form as the Frechet derivative.

Finally, the following form of the mean value theorem for operators is often useful.

Theorem 3.2.1: Let F be continuously Frechet differentiable in a ball Ω_{x_0} containing x_0 in $\mathcal{B}_1$. Then for all u,v in Ω_{x_0} ,

$$F(v) - F(u) = \int_0^1 F'(v + \tau(u - v))(u - v)d\tau$$

where $F'(w)$ denotes the Frechet derivative of F at w .

Proof: It is enough to prove that

$$\frac{d}{d\tau}F(u + \tau(v - u)) = F'(u + \tau(v - u))(v - u) .$$

The result then follows by integration in a Banach space. (We must apply the fundamental theorem of calculus to a Banach space valued function; for a discussion of the integration theory of Banach space valued functions, see Dunford and Schwartz, vol. I)

Since $u + \tau(v - u))$ lies in Ω_{x_0} for $0 \leq \tau \leq 1$

$$\frac{F(u + \tau'(v - u) - F(u + \tau(v - u))}{\tau' - \tau}$$

$$= F'(u + \tau(v - u))(v - u) + \frac{R(u + \tau(v - u) \cdot (\tau - \tau')(v - u))}{\tau - \tau'}$$

But $\|R(z,w)\| = o(\|w\|)$ uniformly for z in Ω_{x_0}, so letting $\tau' \to \tau$ we get the desired result.

3.3 Implicit Function Theorem:

Lemma 3.3.1: Let F be continuously Frechet differentiable in a neighborhood of the origin and let $F'(x) \underset{x \to 0}{\to} 0$; that is, $\lim_{\|x\| \to 0} \|F'(x)\| = 0$. Then there is a continuous function $\varphi(t) \geq 0$, $\varphi(t) \underset{t \to 0}{\to} 0$ such that

$$\|F(x) - F(y)\| \leq \varphi(\max \|x\|, \|y\|) \, \|x - y\| .$$

Proof: By the mean value theorem

$$F(x) - F(y) = \int_0^1 F'(y + \tau(x - y))(x - y)d\tau$$

so

$$\|F(x) - F(y)\| \leq \int_0^1 \|F'(y + \tau(x - y))\| \, \|x - y\| d\tau .$$

Let $\varphi(t) = \sup_{\|x\| \le t} \|F'(x)\|$. Then

$$\|F'(y + \tau(x - y))\| \le \varphi(\max \|x\|, \|y\|)$$

and the conclusion follows.

If F is Frechet differentiable in a neighborhood of the origin, then so is the remainder operator

$$R(x) = F(x) - F(0) - F'(0)x \quad .$$

In fact, $R'(x) = F'(0)$ and $\|R'(x)\| \to 0$ as $\|x\| \to 0$. Applying the lemma to R we have

<u>Corollary 3.3.2</u>: <u>If F is continuously Frechet differentiable in a neighborhood of the origin and if $R(x) = F(x) - F(0) - F'(0)x$, then $\|R(x) - R(y)\| = o(1)\|x - y\|$ as $\|x\|, \|y\| \to 0$.</u>

Now let $\lambda \in R$ and $x \in \mathcal{B}$, and consider a differentiable mapping $F(\lambda,x) : R \times \mathcal{B} \to \mathcal{B}_2$. The Frechet derivative of F is the operator pair (F'_λ, F'_x) mapping $R \times \mathcal{B}$ to $\mathcal{B}_2$. (See exercise 4, p.70)

<u>Theorem (Implicit function theorem)</u>: <u>Let $F(\lambda,x)$ be continuously Frechet differentiable in λ, x in a neighborhood of the origin, and suppose</u>

(i) $F(0,0) = 0$

(ii) $F'_x(0,0)$ has a bounded inverse from $\mathcal{B}_2$ to $\mathcal{B}$.

<u>Then there exists (for small λ) a function $x(\lambda) : R \to \mathcal{B}_1$ such that</u>

(i) $x(0) = 0$

(ii) $x'(\lambda)$ exists and is continuous,

(iii) $F(\lambda, x(\lambda)) \equiv 0$.

<u>Proof</u>: We must solve

$$F(\lambda,x) = F(0,0) + F'_\lambda(0,0)\lambda + F'_x(0,0)x + R(\lambda,x) = 0 ,$$

Noting that $F(0,0) = 0$ and writing $K = [F'_x(0,0)]^{-1}$ this can be written

$$x + KF'_\lambda(0,0)\lambda + KR(\lambda,x) = 0 \quad . \tag{3.3.1}$$

Let $x_0 = 0$ and define a sequence $\{x_n\}$ by

$$x_{n+1} + KF'_\lambda(0,0)\lambda + KR(\lambda,x_n) = 0 \quad .$$

Now R is a continuous mapping from $R \times \mathcal{B}$ to $\mathcal{B}_2$ and K is a continuous mapping from $\mathcal{B}_2$ to $\mathcal{B}$. By Corollary 3.3.2,

$$\|x_{n+1} - x_n\| \le \|K\| \; \|R(\lambda,x_n) - R(\lambda,x_{n-1})\| \; ,$$

$$\le \|K\| \; \varphi(\max \|x_n\|,\|x_{n-1}\|) \; \|x_n - x_{n-1}\| \quad .$$

In particular, $x_1 = \lambda KF'_\lambda(0,0) - KR(\lambda,0)$, and $\|x_1(\lambda)\| \to 0$ as $|\lambda| \to 0$. By choosing λ sufficiently small we can ensure that $\|K\| \; \varphi(\max \|x_n\|,\|x_{n-1}\|) \le \frac{1}{2}$ for $n = 1, 2, \ldots$ and $\|x_{n+1} - x_n\| \le \frac{1}{2} \|x_n - x_{n-1}\|$. By the usual arguments in successive approximation methods we get $\|x_{n+1} - x_n\| \le (\frac{1}{2})^n \; \|x_1 - x_0\| \underset{n \to \infty}{\to} 0$.

It is also easy to see that $x(\lambda)$ is continuous. These steps are left as an exercise (Exercise 5).

To prove differentiability write

$$0 = \frac{F(\lambda,x(\lambda)) - F(\sigma,x(\sigma))}{\lambda - \sigma}$$

$$= F'_{\sigma}(\sigma,x(\sigma)) + F'_{x}(\sigma,x(\sigma)) \frac{x(\lambda) - x(\sigma)}{\lambda - \sigma}$$

$$+ \frac{R(\sigma,x(\sigma);\ \lambda - \sigma,\ x(\lambda) - x(\sigma))}{(\lambda - \sigma)}$$

By the properties of the remainder operator (since $x(\lambda)$ is continuous),

$$\lim_{\lambda \to \sigma} \frac{\|R(\sigma,x(\sigma);\ \lambda - \sigma,\ x(\lambda) - x(\sigma))\|}{|\lambda - \sigma|} = 0\ , \qquad (3.3.2)$$

so taking limits we get

$$F'_{x}(\sigma,x(\sigma))x'(\sigma) + F'_{\sigma}(\sigma,x(\sigma)) = 0 \quad .$$

For σ near zero, $F'_{x}(\sigma,x(\sigma))$ is invertible, so $x'(\sigma)$ is in fact defined.

Note: To apply the argument in (3.3.2) we need to know that $\|x(\lambda) - x(\sigma)\| = O(|\lambda - \sigma|)$. But from (3.3.1) we get

$$x(\lambda) - x(\sigma) = -KF'_{\lambda}(0,0)(\lambda - \sigma) - K(R(\lambda,x(\lambda)) - R(\sigma,x(\sigma)))$$

hence

$$\|x(\lambda) - x(\sigma)\| \le c_1|\lambda - \sigma| + o(|\lambda - \sigma|) + o(\|x(\lambda) - x(\sigma)\|)\ ,$$

and the desired result follows.

3.4 Analyticity.

A mapping $x(\lambda) : C \to B$ where C denotes the complex numbers and B denotes a Banach space is said to be analytic in $|\lambda - \lambda_0| < \delta$ if there is a power series

$$x(\lambda) = \sum_{k=0}^{\infty} x_k(\lambda - \lambda_0)^k$$

with coefficients x_k in B which is convergent for $|\lambda - \lambda_0| < \delta$. By convergence we mean that the partial sums

$$S_{m,n} = \sum_{k=m+1}^{n} x_k(\lambda - \lambda_0)^k$$

form a Cauchy sequence in B for $|\lambda - \lambda_0| < \delta$.

An operator $A : B_1 \to B_2$ is analytic in $\|x - x_0\| < \delta$ if there is a sequence of bounded homogeneous operators of degree k such that the series

$$A(x_0 + v) = \sum_{k=0}^{\infty} A_k(v) \tag{3.4.1}$$

is convergent for $\|v\| < \delta$. By A_0 we understand a fixed vector in B_2, while $A_k(\lambda v) = \lambda^k A_k(v)$. When we say A_k is bounded we mean that

$$M_k = \sup_{\|x\|_1=1} \|A_k(x)\|_2 < +\infty$$

It follows that $\|A_k(x)\| \leq M_k\|x\|^k$ for all $x \in B_1$.

It turns out that the operator A_k in (3.4.1) comes from a k-linear operator, as we shall see below.

Clearly, if $x(\lambda)$ is analytic in the above sense for $|\lambda - \lambda_0| < \delta$, then $\langle x(\lambda),\varphi^*\rangle$ is analytic in $|\lambda - \lambda_0| < \delta$ in the usual sense, for any $\varphi^* \in B^*$. Similarly, if A is analytic in $\|x - x_0\| < \delta$, then for $\|v\| < 1$, $\langle A(x_0 + \lambda v),\varphi^*\rangle$ is analytic in $|\lambda| < \delta$ in the usual sense for any φ^* in B^*. The converses of thes statements are also valid.

<u>Theorem 3.4.1</u>: <u>A necessary and sufficient condition that a function $x(\lambda)$ or an operator $A(x)$ be analytic in $|\lambda| < \delta$ is that $\langle x(\lambda),\varphi^*\rangle$, respectively $\langle A(\lambda x),\varphi^*\rangle$ (where $\|x\| < 1$) be analytic in $|\lambda| < \delta$ in the usual sense, for any φ^* in B^*</u>.

<u>Proof</u>: First consider the case where $\langle x(\lambda),\varphi^*\rangle$ is analytic in λ. Then

$$\langle x(\lambda),\varphi^*\rangle = \frac{1}{2\pi i}\int_C \frac{\langle x(t),\varphi^*\rangle}{t - \lambda}\,dt \quad .$$

Now the integral

$$\frac{1}{2\pi i}\int_C \frac{x(t)}{t - \lambda}\,dt$$

is defined as the strong limit of a sequence of finite sums, so it is permissable to interchange the integration with the functional operation $\langle\, ,\varphi^*\rangle$: We get

$$\left\langle x(\lambda) - \frac{1}{2\pi i}\int_C \frac{x(t)}{t - \lambda}\,dt,\ \varphi^*\right\rangle = 0$$

for all φ^* in B^*. This means, however, that

$$x(\lambda) = \frac{1}{2\pi i} \int_C \frac{x(t)dt}{t - \lambda} .$$

From this representation we obtain the convergent power series in λ in the usual way:

$$x(\lambda) = \sum_{k=0}^{\infty} x_k \lambda^k$$

where

$$x_k = \frac{1}{2\pi i} \int_C \frac{x(t)}{t^{k+1}} dt .$$

In the case of an operator we obtain, as above

$$A(\lambda x) = \sum_{k=0}^{\infty} \lambda^k A_k(x) \tag{3.4.2}$$

where

$$A_k(x) = \frac{1}{2\pi i} \int_C \frac{A(tx)}{t^{k+1}} dt . \tag{3.4.3}$$

Here $\|x\| \leq 1$ and C is the circle of radius r, $|\lambda| < r < \delta$. The A_k's are easily seen to be homogeneous of degree k and bounded. (Boundedness of A_k follows from continuity at $x = 0$ which in turn follows from the continuity of A).

For $\|x\| < \delta$ we may set $\lambda = 1$ in (3.4.2) and obtain the desired series representation for A .

Corollary 3.4.2: Let A be an operator on a complex Banach space (a Banach space over the complex numbers) and let A be Frechet differentiable everywhere in the neighborhood of a point $x_0 \in B_1$. Then A is analytic there.

Proof: If A is Frechet differentiable in the set $\Omega_{x_0} = \{u : \|x_0 - u\| < \delta\}$ then the function $\langle A(x_0 + \lambda v), \varphi^* \rangle$ is differentiable as a function of λ in $|\lambda| < \delta$ (where $\|v\| < 1$) for any φ^*. The conclusion now follows from Theorem 3.4.1. Note that the radius of convergence of the power series for $A(x_0 + \lambda v)$ is at least δ (when $\|v\| \leq 1$).

Corollary 3.4.3: If the operator $F(\lambda, x)$ in the implicit function theorem is Frechet differentiable in x and λ for λ complex and x in a complex Banach space then the function $x(\lambda)$ obtained in the implicit function theorem is analytic.

Finally, let us show that the operators defined by (3.4.3) come from multilinear operators. Since $A(x)$ is analytic (say) for $\|x\| < \delta$, it is immediate that $\langle A(\lambda_1 x_1 + \ldots + \lambda_n x_n), \varphi^* \rangle$ is analytic in $\lambda_1, \ldots, \lambda_n$ provided $\|\lambda_1 x_1 + \ldots + \lambda_n x_n\| < \delta$. We can thus get the representation

$$A(\lambda_1 x_1 + \ldots + \lambda_n x_n) = \frac{1}{(2\pi i)^n} \int_C \cdots \int_C \frac{A(t_1 x_1 + \ldots + t_n x_n)}{(t_1 - \lambda_1)\ldots(t_n - \lambda_n)} \, dt_1 \ldots dt_n$$

Where C is a circle of suitably small radius. The above representation need only hold for suitably small $\lambda_1, \ldots, \lambda_n$, $\|x_1\|, \ldots, \|x_n\|$.

Now set

$$B_n(x_1, \ldots, x_n) = \frac{\partial^n}{\partial\lambda_1 \ldots \partial\lambda_n} A(\lambda_1 x_1 + \ldots + \lambda_n x_n)\Big|_{\lambda_1 = \ldots = \lambda_n = 0} .$$

We claim that (i) B_n is continuous and linear in each of its arguments. (ii) $B_n(x, \ldots x) = n!\, A_n(x)$, where A_n is given by (3.4.3).

That B_n is continuous follows from the representation

$$B_n(x_1, \ldots x_n) = \frac{1}{(2\pi i)^n} \int \ldots \int \frac{A(t_1 x_1 + \ldots + t_n x_n)}{t_1^2 \ldots t_n^2}\, dt_1 \ldots dt_n$$

and the fact that A is continuous on $\|x\| < \delta$. (We only need to establish continuity of B in a neighborhood of the origin. Why?) To prove linearity, write

$$B_n(x_1 + y_1, x_2, \ldots x_n) = \frac{\partial^n}{\partial\lambda_1 \ldots \partial\lambda_n} A(\lambda_1(x_1 + y_1) + \ldots + \lambda_n x_n)\Big|_{\lambda_i = 0} .$$

Now use the fact that for a function $F(s,t)$,

$$\frac{d}{d\lambda} F(\lambda,\lambda)\Big|_{\lambda = 0} = \frac{\partial F}{\partial s}(s,0)\Big|_{s = 0} + \frac{\partial F}{\partial t}(0,t)\Big|_{t = 0} .$$

Then we get

$$B_n(x_1 + y_1, x_2, \ldots x_n) = \frac{\partial^n}{\partial s \partial \lambda_2 \ldots \partial \lambda_n} A(sx_1 + \lambda_2 x_2 + \ldots + \lambda_n \lambda_n) \Big|_{s = \lambda_2 = \ldots = 0}$$

$$+ \frac{\partial^n}{\partial t \partial \lambda_2 \ldots \partial \lambda_n} A(ty_1 + \lambda_2 x_2 + \ldots + \lambda_n x_n) \Big|_{t = \lambda_2 = \ldots = 0}$$

$$= B_n(x_1, x_2, \ldots x_n) + B(y_1, x_2, \ldots x_n) \quad .$$

The homogeneity relation $B(\lambda x_1, x_2, \ldots x_n) = \lambda B(x_1, x_2, \ldots x_n)$ is proved as follows:

$$B_n(\lambda x_1, x_2, \ldots x_n) = \frac{\partial^n}{\partial \lambda_1 \ldots \partial \lambda_n} A(\lambda_1 \lambda x_1 + \ldots + \lambda_n x_n) \Big|_{\lambda_i = 0}$$

$$= \frac{\lambda \, \partial^n}{\partial s \partial \lambda_2 \ldots \partial \lambda_n} A(sx_1 + \ldots + \lambda_n x_n) \Big|_{s = \lambda_2 = \ldots = 0}$$

$$= \lambda B_n(x_1, \ldots x_n)$$

Now let's compute $B_n(x, \dots x)$. We get

$$B_n(x, \dots x) = \frac{\partial^n}{\partial\lambda_1 \dots \partial\lambda_n} A(\lambda_1 x + \dots + \lambda_n x)\Big|_{\lambda_i = 0}$$

$$= \frac{\partial^n}{\partial\lambda_1 \dots \partial\lambda_n} A(\lambda x)\Big|_{\lambda_i = 0}$$

(where $\lambda = \lambda_1 + \dots + \lambda_n$)

$$= \frac{d}{d\lambda}^n A(\lambda x)\Big|_{\lambda = 0} \quad .$$

The last step follows from the rule $\frac{\partial}{\partial\lambda_i} = \frac{d}{d\lambda}\frac{\partial\lambda}{\partial\lambda_i} = \frac{\partial}{\partial\lambda}$. On the other hand, from (3.4.2),

$$A(\lambda x) = \sum_{n=0}^{\infty} \lambda^n A_n(x)$$

so

$$\frac{d}{d\lambda}^n A(\lambda x)\Big|_{\lambda = 0} = \quad n!\ A_n(x) \quad . \qquad \text{Q.E.D.}$$

Exercises:

1) Show

(i) $A_k(x)$ defined by (3.4.2) is homogeneous of degree k .

(ii) $A_k(x)$ is defined and continuous for all x .

(iii) The constant

$$A_k = \sup_{\|x\|=1} \|A_k(x)\|$$

is finite

(iv) $\|A_k(x)\| \leq A_k\|x\|^k$.

2) Let $x(\lambda) : \mathbb{R} \to B$. Show that if $x(\lambda)$ is n times continuously differentiable then Taylor's theorem holds:

$$x(\lambda) = x(\lambda_0) + x'(\lambda_0)(\lambda - \lambda_0) + \ldots + \\ + \frac{x^n(\lambda_0)(\lambda - \lambda_0)^n}{n!} + o(|\lambda - \lambda_0|^n)$$

3) Let F be a mapping from $\mathcal{B}_1$ to $\mathcal{B}_2$ that is n times continuously Frechet differentiable. Show that the operators B_k given in the definition of the Taylor series are given by

$$B_n(x_1, \ldots x_k) = \frac{\partial^k}{\partial\lambda_1 \ldots \partial\lambda_k} F(x_0 + \lambda_1 x_1 + \ldots + \lambda_k x_k)\Bigg|_{\lambda_i = 0} .$$

4) Let $F(\lambda,x)$ be a mapping from $R \times B$ into B and suppose that the partial Frechet derivatives $F_\lambda(\lambda,x)$ (x fixed) and $F_x(\lambda,x)$ (λ fixed) exist and are continuous in a domain Ω . Prove that the total Frechet derivative of F exists and is (F_λ,F_x) . Generalize to mappings $F : A \times B \to C \times D$.

5) Complete the convergence arguments in the proof of the implicit function theorem. Hint: choose δ so that $0 < \delta < 1$ and $\varphi(\delta) \leq 1/2\|K\|$. Then choose λ so small that $\|x_1(\lambda)\| \leq \delta/2$. Show that $x(\lambda)$ is continuous.

3.5 Decomposition of Vector Fields.

The following decomposition of vector fields is importent in the study of the Navier-Stokes equations. Let Ω be a smoothly bounded domain in R^3. Let $L_2(\Omega)$ be the Hilbert space of vector fields

$$L_2 = \{\underline{u} : \underline{u} = (u_1,u_2,u_3) \ , \int_\Omega u_i u_i d\underline{x} < +\infty\}$$

with inner product

$$(\underline{u},\underline{v}) = \int_\Omega u_i v_i d\underline{x} \ .$$

The subspace H_σ is defined to be the sub-space of vector fields $\underline{w}$ for which

$$\int_\Omega \underline{w} \cdot \nabla\varphi \, d\underline{x} = 0$$

for all φ in $C^1(\bar{\Omega})$.

If $\underline{w}$ is smooth then by Green's theorem

$$0 = \int_{\partial\Omega} \varphi \, w_\nu d\sigma - \int_\Omega \varphi \ \mathrm{div}\ \underline{w}\ d\underline{x} \ . \tag{3.5.1}$$

Since φ is arbitrary we first let φ range over continuous functions which vanish on $\partial\Omega$. Then we see that

$$\int_\Omega \varphi \ \mathrm{div}\ \underline{w}\ d\underline{x} = 0 \ ,$$

for all continuous φ vanishing on $\partial\Omega$, so that $\text{div}\, \underline{w} = 0$ in Ω. Now going back to (3.5.1) we get

$$\int_{\partial\Omega} \varphi\, w_\nu d\sigma = 0 \tag{3.5.2}$$

for arbitrary φ. This condition implies that $w_\nu = \underline{w} \cdot \nu = 0$ on $\partial\Omega$. Conversely, if $\underline{w}$ is smooth, $\text{div}\, \underline{w} = 0$, and $w_\nu = 0$ on $\partial\Omega$, then $w \in H_\sigma$.

It is easily seen that H_σ is a closed subspace of $L_2(\Omega)$. We let H_π denote the orthogonal complement of H_σ.

We now wish to prove that if $\underline{u}$ is any $C_{1+\alpha}$ vector field on Ω, then

$$\underline{u} = \underline{w} + \nabla p$$

where $\underline{w} \in H_\sigma$ and $\nabla p \in H_\pi$.

First, we solve the Neumann problem

$$\Delta p = \text{div}\, \underline{u}\,, \qquad \frac{\partial p}{\partial \nu} = \underline{u} \cdot \underline{\nu}\,. \tag{3.5.3}$$

Then put $\underline{w} = \underline{u} - \nabla p$. By the Schauder estimates for elliptic boundary value problems, $\underline{u} \in C_{1+\alpha}$ implies $p \in C_{3+\alpha}$ and $\nabla p \in C_{2+\alpha}$. So $\text{div}\, \underline{w} = \text{div}\, \underline{u} - \Delta p = 0$ and $\underline{w} \cdot \nu = \underline{u} \cdot \nu - \frac{\partial p}{\partial \nu} = 0$ on $\partial\Omega$. Thus $\underline{w}$ is a $C_{1+\alpha}$ vector field in H_σ.

We can get the following representation for p: If $G(x,y)$ is the Neumann function for the Laplacian on Ω, then by the usual arguments of potential theory

$$p(x) = \int_\Omega \nabla_y\, G(x,y)\underline{u}(y)d\underline{y} \tag{3.5.4}$$

(the boundary terms have cancelled)

From the representation (3.5.4) it can be shown that the transformation $u \to p$ satisfies $|p|_{k+\alpha+1} \le C|u|_{\alpha+k}$ for any integer $k = 1,2, \ldots$, and some constant C depending only on α and k. Therefore $|\nabla p|_{k+\alpha} \le C|u|_{k+\alpha}$. Furthermore, the decomposition $\underline{u} = \underline{w} + \nabla p$ is unique (up to an additive constant in p). Denote by P_π the transformation $P_\pi \underline{u} = \nabla p$. Then $|P_\pi u|_{k+\alpha} \le c|u|_{k+\alpha}$, $k = 1,2, \ldots$. Moreover $P_\pi^2 = P_\pi$ on C_α. In fact, if $\underline{u} = \underline{w} + \nabla p$ then $P_\pi \underline{u} = \nabla p$. If we decompose ∇p as $\nabla p = \underline{w}' + \nabla\varphi$ we see that $\underline{w}' = 0$ and $\nabla p = \nabla\varphi$; therefore $P_\pi^2 \underline{u} = P_\pi \nabla p = \nabla p = P_\pi \underline{u}$, so $P_\pi^2 = P_\pi$, and P_π is a projection. Define $P_\sigma = I - P_\pi$. Then P_σ is a projection,

If $\underline{u}$ is a smooth vector field then $(P_\sigma \underline{u}, P_\pi \underline{u}) = 0$, so $\|\underline{u}\|^2 = \|P_\sigma \underline{u}\|^2 + \|P_\pi \underline{u}\|^2$ and $\|P_\sigma \underline{u}\| \le \|\underline{u}\|$, $\|P_\pi \underline{u}\| \le \|\underline{u}\|$ where $\| \;\|$ denotes the L_2 norm in $L_2(D)$. Therefore P_σ and P_π are bounded operators in the L_2 norm and can be extended to all of $L_2(\mathcal{D})$ by continuity. We thus have orthogonal projections P_σ and P_π on $L_2(\mathcal{D})$.

In order to avoid using the representation for the pressure given by (3.5.4) we proceed as follows. Given the vector field u in $\mathcal{D}$, extend u beyond $\mathcal{D}$ to a function with compact support in R^3. Call this extension $\tilde{u}$. $\tilde{u}$ can be chosen so that $|\tilde{u}|_{1+\alpha} \le c|u|_{1+\alpha}$ for some constant c independent of u. Put

$$p_1(x) = -\frac{1}{4\pi}\int_{R^3} \frac{1}{|x-y|} \operatorname{div} \tilde{u}\, dy = \frac{1}{4\pi}\int_{R^3} \nabla_y \frac{1}{|x-y|} \cdot \tilde{u}(y)dy .$$

Then $\nabla p_1 = \tilde{u}(x)$. This can be seen, for example, by Fourier transforms: If $\hat{p}_1(\xi)$ denotes the Fourier transform of p_1, then

$$\widehat{(\operatorname{div} \tilde{u})} = i\, \xi_j\, \hat{u}_j \qquad \text{(sum over } j \text{)}$$

and $\hat{p}_1(\xi) = \frac{1}{|\xi|^2}$ (div $\hat{u}$) $= i \frac{\xi_j \hat{u}_j}{|\xi|^2}$. Therefore

$$\frac{\partial p_1}{\partial x_k} = i \, \xi_k \, \hat{p}_1 = \frac{\xi_k \xi_j \tilde{u}_j}{|\xi|^2} \quad .$$

But $- \frac{\xi_k \xi_j}{|\xi|^2}$ is the Fourier transform of the singular (Calderon-Zygmund) kernel $\frac{\partial}{\partial x_k} \frac{\partial}{\partial x_j} \frac{1}{4\pi |x - y|}$, so the result follows.

The classical Schauder estimates for the singular integral operator

$$\nabla p_1 = \frac{1}{4\pi} \nabla_x \int_{R^3} \nabla_y \frac{1}{|x - y|} \cdot \tilde{u}(y) \, dy$$

show that $|\nabla p_1|_{k+\alpha} \leq c |\tilde{u}|_{k-2+\alpha}$ for any positive integer k .

It remains to satisfy the boundary conditions. Accordingly, it is necessary to add a correction term to p_1 , so that $p = p_1 + p_2$ satisfies $(\underline{u} - \nabla p) \cdot \nu = 0$ on the boundary. This leads to the Neumann problem

$$\Delta p_2 = 0$$

$$\frac{\partial p_2}{\partial \nu} = u \cdot \nu - \frac{\partial p_1}{\partial \nu} \quad .$$

The solvability condition is automatically satisfied:

$$\int_{\partial \mathcal{D}} (u \cdot \nu - \frac{\partial p_1}{\partial \nu}) d\sigma = \int_{\mathcal{D}} \text{div } u \, dx - \int_{\mathcal{D}} \Delta p_1 \, dx = 0$$

If u is $C_{1+\alpha}$ and $\partial \mathcal{D}$ is $C_{2+\alpha}$ then p_1 is in $C_{2+\alpha}$, and $\frac{\partial p_1}{\partial \nu}$

is in $C_{1+\alpha}$, with $\left|\frac{\partial p_1}{\partial \nu}\right|_{1+\alpha} \leq c|u|_{1+\alpha}$. The problem is now reduced to the classical Schauder estimates for the Neumann problem.

Exercises

1. Show that a necessary condition that the Neumann problem

$$\Delta\varphi = f\ , \quad \frac{\partial\varphi}{\partial\nu}\Big|_{\partial\Omega} = g$$

be solvable is

$$\int_{\Omega} f\, d\underline{x} = \int_{\partial\Omega} g\, d\sigma \quad .$$

Is the solution unique? This condition is also sufficient. Show it is satisfied for the problem (3.5.3).

2. Prove that if $\underline{u} \in C_{1+\alpha}$ then the decomposition $\underline{u} = \underline{w} + \nabla p$ is unique (up to an additive constant factor in p).

3. Show that if $\operatorname{div} \underline{w} = 0$ and $\underline{w} \cdot \nu = 0$ on ∂D then $\underline{w}$ is orthogonal to all gradients in the Hilbert space $L_2(D)$.
Show that if $P^2 = P$ and $Q = I - P$ then $Q^2 = Q$. Show that P_π and P_σ are orthogonal projections onto H_π and H_σ. That is, $(P_\sigma \underline{u}, P_\pi \underline{u}) = 0$ for all $\underline{u}$ in $L_2(D)$.

Notes

The Riesz theory for compact linear operators on a Hilbert Space is developed in Riesz-Nagy, "Functional Analysis", secs. 77 - 80. The modifications necessary for the extension of the theory to general Banach spaces are outlined in sec. 89. The solvability of the Neumann problem is discussed briefly on pp. 190-194.

For further discussion of the implicit function theorem see the review article by Vainberg and Trenogin.

See Ladyzhenskaya's book, "The Mathematical Theory of Viscous Incompressible Flow" for a detailed discussion of vector fields, their decomposition, and properties of the projections P_σ and P_π .

IV

Bifurcation At A Simple Eigenvalue

In this chapter we treat the bifurcation problem and the stability of bifurcating solutions at a simple eigenvalue. We treat the Navier-Stokes equations as a specific, but important physical case, which will exhibit many of the techniques needed to discuss the problem under more general circumstances.

4.1 The Navier-Stokes Equations.

The (time independent) Navier-Stokes equations in Cartesian coordinates are

$$\begin{aligned} \Delta u_i &= \frac{\partial p}{\partial x_i} + \lambda u_j \frac{\partial u_i}{\partial x_j} \\ \frac{\partial u_i}{\partial x_i} &= 0 \end{aligned} \tag{4.1.1}$$

Here p is the pressure of the fluid, u_i are the components of the velocity of the fluid along the x_1, x_2, and x_3 axes; repeated indices denote summation, and λ is a parameter. We shall at times drop the subscripts and write u for the vector field. Equations (4.1.1) are to be considered in a bounded domain $\mathcal{D}$ which remains fixed in time but whose boundaries move uniformly in a tangential direction; we may also allow fluid to flow in and out of $\mathcal{D}$ across the boundaries. The fluid, due to viscosity, adheres to the (moving) walls, so we must have the boundary conditions

$$u_i \Big|_{\partial D} = \psi_i \tag{4.1.2}$$

Due to the divergence-free equation $(\operatorname{div} u = 0)$, the boundary conditions satisfy a condition of no net flux:

$$\int_{\partial\mathcal{D}} \psi_\nu d\sigma = \int_{\partial\mathcal{D}} u_i \nu_i d\sigma = \int_{\partial\mathcal{D}} \frac{\partial u_i}{\partial x_i} dx = 0 \tag{4.1.3}$$

If the components ψ_i satisfy condition (4.1.3) then the boundary value problem (4.1.1 - 4.1.2) has at least one solution for all values of λ ! This theorem, due to Leray, will be proved later, using topological degree methods.

Suppose for now, however, that a solution (u_0, λ_0) of equations (4.1.1) is known. We wish to investigate the nearby solutions for λ near λ_0. We look for them in the form $u(\lambda) = u_0 + v$, where $\operatorname{div} v = 0$ and $v = 0$ on $\partial\mathcal{D}$ (since u_0 satisfies the boundary conditions).

Operating on equation (4.1.1) by the projection P_σ we may write it in the form

$$Au_0 = \lambda_0 N(u_0, u_0)$$

where $Au = P_\sigma \Delta u$ and $N(u_0, u_0) = P_\sigma u_0 \cdot \nabla u_0$. Writing $u = u_0 + v$ we get for v the equation

$$A(u_0 + v) = \lambda N(u_0 + v,\ u_0 + v) \quad ,$$

which can be written in the form

$$L_0 v = \tau N(u_0,u_0) + \lambda N(v,v) + \tau M(u_0,v) \tag{4.1.4}$$

where

$$\lambda = \lambda_0 + \tau$$

$$L_0 v = Av - \lambda_0 M(u_0,v) \quad ,$$

$$M(u_0,v) = N(u_0,v) + N(v,u_0) \quad .$$

Let $C_{k+\alpha,\sigma}$ denote the Banach space of vector fields u such that $\operatorname{div} u = 0$ and $|u_i|_{k+\alpha} < +\infty$. We assume $u_0 \in C_{2+\alpha,\sigma}$ and look for solutions of (4.1.4) in that class. There are two cases depending on whether L_0 is invertible or not. Note that L_0 is the derived operator at the known solution (u_0,λ_0). In this section we consider the case where L_0 is invertible (no bifurcation).

<u>Lemma 4.1.1</u>: <u>L_0 is invertible if $L_0\varphi = 0$ has no nontrivial solutions; and then there is a constant C such that</u>

$$|L_0^{-1}u|_{2+2\alpha} \leq C|u|_\alpha \quad . \tag{4.1.5}$$

Lemma 4.1.1 will be proved below. We shall need the inequality

$$|N(u,v)|_\alpha \leq C|u|_\alpha |v|_{1+\alpha} \tag{4.1.6}$$

which is a consequence of the multiplicative property of the Hölder norms (exercise 4, p.54). The proof of (4.1.6) is left as an exercise.

<u>Theorem 4.1.2</u>: <u>If L_0 is invertible then (4.1.4) has a solution v analytic in τ for small τ. ($v \in C_{2+\alpha,\sigma}$)</u>

Proof: Operating on (4.1.4) by L_0^{-1}, we get

$$v - \tau L_0^{-1}N(u_0,u_0) - \lambda L_0^{-1}N(v,v) - \tau L_0^{-1}M(u_0,v) = 0 \qquad (4.1.7)$$

From (4.1.5) and (4.1.6) we see that $|L_0^{-1}N(u,u)|_{2+\alpha} \leq C(|u|_{1+\alpha})^2$. Therefore $L_0^{-1}N(u,u)$ is an analytic mapping from $C_{2+\alpha,\sigma}$ to itself. (It is even completely continuous). Equation (4.1.7) is a functional equation for v in the form $\mathcal{F}(v,\tau) = 0$. We have $\mathcal{F}(0,0) = 0$ and $\mathcal{F}_v(0,0) = I$ (the identity). Therefore by the implicit function theorem there exists an analytic solution $v(\tau)$ of (4.1.7) which converges in the Banach space $C_{2+\alpha,\sigma}$.

It remains to prove Lemma 4.1.1. We first need the following lemma, due to Solonnikov: (See Ladyzhenskaya, pp. 78-79)

Lemma 4.1.3: Let $f_i \in C_\alpha$ and let u_i, p be the solutions of

$$\Delta u_i = \frac{\partial p}{\partial x_i} + f_i$$

$$\frac{\partial u_i}{\partial x_i} = 0 \qquad (4.1.8)$$

$$u_i \nu_i = 0 \quad \text{on} \quad \partial\Omega$$

Then there is a constant c such that $|u|_{2+\alpha} \leq c|f|_\alpha$ and $|p|_{1+\alpha} \leq c|f|_\alpha$. The proof of this lemma is carried out using hydrodynamic potentials.

Proof of Lemma 4.1.1: Lemma 4.1.3 shows that A^{-1} is a bounded transformation from $C_{\alpha,\sigma}$ to $C_{2+\alpha,\sigma}$. Operating on $L_0 u = f$ by A^{-1} we get

$$v - \lambda_0 A^{-1}M(u_0,v) = A^{-1}f \qquad (4.1.9)$$

It is easily seen that $|M(u_0,v)|_\alpha \le c|v|_{1+\alpha}$ and hence that $A^{-1}M(u_0,v)$ is a bounded transformation from $C_{\sigma,1+\alpha}$ to $C_{\sigma,2+\alpha}$. This makes it a compact operator from $C_{\sigma,2+\alpha}$ to itself, so we may apply the Fredholm alternative to the functional equation (4.1.9). Either (4.1.9) is boundedly invertible for all f, and $|v|_{2+\alpha} \le c|A^{-1}f|_{2+\alpha}$; or else the homogeneous equation $(f = 0)$ has a non-trivial solution. In that case, however, we have

$$v - \lambda_0 A^{-1}M(u_0,v) = 0$$

hence

$$Av - \lambda_0 M(u_0,v) = 0$$

which means that $L_0 v = 0$. Thus if $L_0\varphi = 0$ has only the trivial solution, then (4.1.9) is in fact boundedly invertible, and the conclusion of Lemma 4.1.1 follows.

Exercise:

1. Show that the equations

$$\Delta u_i = u_j \partial_j u_i + \frac{\partial p}{\partial x_i} + f_i \qquad \frac{\partial u_i}{\partial x_i} = 0$$

and the functional equation

$$u = A^{-1}N(u,u) + A^{-1}P_\sigma f$$

in $C_{2+\alpha,\sigma}$ are equivalent. That is, a solution of one problem leads to a solution of the other.

4.2 Bifurcation

We wish to study solutions of (4.1.1) in a neighborhood of a known solution (λ_0,u_0) when the operator L_0 is not invertible - specifically, when 0 is a simple eigenvalue of L_0. It will be convenient to set up the problem in the following way:

Choose w so that $\operatorname{div} w = 0$,

$$\Delta w_i = \frac{\partial q}{\partial x_i}$$

(This problem is solvable, and if $\partial\mathcal{D}$ and ψ are of class $C_{2+\alpha}$, then $w \in C_{\sigma,2+\alpha}(\bar{\mathcal{D}}))$. Write solutions of (4.1.1) in the form $u = w + v$. For v we get the equations

$$\Delta v_i = \frac{\partial p}{\partial x_i} + \lambda w_j \frac{\partial w_i}{\partial x_j} + \lambda\left(w_j \frac{\partial v_i}{\partial x_j} + v_j \frac{\partial w_i}{\partial x_j}\right)$$

$$+ \lambda v_j \frac{\partial v_i}{\partial x_j} \; .$$

$$\frac{\partial v_i}{\partial x_i} = 0 \qquad v_i\Big|_{\partial\mathcal{D}} = 0$$

This equation may be written in functional form

$$Av - \lambda Bv - \lambda N(v,v) = \lambda f \tag{4.2.1}$$

where

$$f = P_\sigma w \cdot \nabla w$$

$$Bv = P_\sigma[w \cdot \nabla v + v \cdot \nabla w] \; .$$

Let us suppose that $v(\lambda)$ is a solution, analytic in λ for $|\lambda - \lambda_0|$ sufficiently small, and look for other solutions of (4.2.1) in the form $u = v + z$. For z we get

$$(A - \lambda B)(v + z) - \lambda N(v + z, v + z) = \lambda f \; ,$$

which reduces, after some cancellation, to the equation

$$L(\lambda)z - \lambda N(z,z) = 0 \tag{4.2.2}$$

where

$$L(\lambda) = A - \lambda B - \lambda M(v(\lambda),\ .)$$

One solution of equation (4.2.2) is $z \equiv 0$, corresponding to the original solution v. We shall construct non-trivial solutions of (4.2.2) in the neighborhood of points $(\lambda_0, v(\lambda_0))$ for which the operator $L_0 = L(\lambda_0)$ has 0 as a simple eigenvalue. Since $v(\lambda)$ is <u>assumed</u> analytic in λ for λ near λ_0 (this happens in a number of physical cases), $L(\lambda)$ is an analytic operator ($L(\lambda)$ mapping $C_{\sigma,2+\alpha}$ to $C_{\sigma,\alpha}$).

Accordingly, equation (4.2.2) can be written in the form

$$L_0 z + (\tau L_1 + \ldots)z - (\lambda_0 + \tau)N(z,z) = 0 \quad .$$

where the dots denote terms of higher order in τ. $(\tau = \lambda - \lambda_0)$

Let $\mu(\lambda)$ be the eigenvalue of $L(\lambda)$ which vanishes at $\lambda = \lambda_0$.

We assume

$$\mu'(\lambda_0) \neq 0 \quad . \tag{4.2.3}$$

The operator L_0 plays an important role in the bifurcation problem. We need the results of the following lemma, which essentially constitute a statement of the Riesz-Schauder theory for L_0.

<u>Lemma 4.2.1</u>: <u>If 0 is a simple eigenvalue of L_0 in H_σ, then $L_0\varphi_0 = 0$, where $\varphi_0 \in H_\sigma$; φ_0 is in fact smooth $(\varphi_0 \in C_{\sigma,2+\alpha})$. Furthermore, there exists an adjoint eigenfunction φ_0^* of L_0^*: $L_0^*\varphi_0^* = 0$, and $(\varphi_0, \varphi_0^*) = 1$. $\varphi_0^* \in C_{\sigma,2+\alpha}$ also. The</u>

operator $Pu = (u,\varphi_0^*)\varphi_0$ is a projection onto the null space of L_0 and $Q = I - P$ is a projection onto the range. P and Q are bounded projections on $C_{\sigma,2+\alpha}$ as well, and there is a bounded transformation $K : C_{\sigma,\alpha} \to C_{\sigma,2+\alpha}$ such that $KL_0 = Q$.

The proof of lemma 4.2.1 will be given later.

Lemma 4.2.2: The critical eigenvalue $\mu(\lambda)$ is analytic in λ and

$$\mu'(\lambda_0) = (L_1\varphi_0,\varphi_0^*)$$

where $L_1\varphi = -B\varphi - \lambda_0 M(v_1,\varphi) - M(v_0,\varphi)$. Here $v(\lambda) = v(\lambda_0) + \tau v_1 + \tau^2 v_2 + \ldots$, where $\tau = (\lambda - \lambda_0)$.

Lemma 4.2.2 is a perturbation result which can be obtained by calculating the power series for the critical eigenvalue $\mu(\lambda)$. (Exercise 1, p. 87)

From now on we shall write

$$[u] = (u,\varphi_0^*) ;$$

then (4.2.3) is equivalent to

$$[L_1\varphi_0] \neq 0 \tag{4.2.4}$$

Note that $[\varphi_0] = 1$, and that $Pw = [w]\varphi_0$. Also, f is in the range of L_0 if and only if $[f] = 0$, so $[Qw] = 0$ for any w.

Theorem 4.2.3: Under the assumption (4.2.4) equation (4.2.2) has a nontrivial solution branch (λ,z) which may be expressed parametrically in the form $z = z(\epsilon)$, $\lambda = \lambda(\epsilon)$, where $z(0) = 0$ and $\lambda(0) = \lambda_0$. The functions z and λ are power series in ϵ, convergent for sufficiently small $|\epsilon|$, and $\epsilon = [z(\epsilon)]$.

Proof: Putting $z = \epsilon w$, $\tau = \epsilon \sigma$, and substituting into (4.2.2), we get

$$L(\lambda_0 + \epsilon \sigma)w - (\lambda_0 + \epsilon \sigma) \epsilon N(w,w) = 0 ,$$

$$L_0 w + \epsilon \sigma L_1 w + \dots - (\lambda_0 + \epsilon \sigma) \epsilon N(w,w) = 0 ,$$

where the dots denote higher order terms in ϵ .
Now write $w = Pw + Qw$. We normalize w so that $[w] = 1$, in which case $Pw = \varphi_0$, and write $Qw = \xi$ so that $w = \varphi_0 + \xi$. Then we get

$$L_0 \xi + \{\epsilon \sigma L_1(\varphi_0 + \xi) + \dots - (\lambda_0 + \epsilon \sigma) \epsilon N(\varphi_0 + \xi, \varphi_0 + \xi)\} = 0 \tag{4.2.5}$$

If (4.2.5) is valid, then the quantity in braces satisfies $[\{ \}] = 0$; hence

$$\sigma[L_1 \varphi_0] + \sigma[L_1 \xi] + \dots - (\lambda_0 + \epsilon \sigma)[N(\varphi_0 + \xi, \varphi_0 + \xi)] = 0 \tag{4.2.6}$$

(Note that we have divided out a factor of ϵ)
Operating on (4.2.5) with K we get

$$\xi + \epsilon K\{\sigma L_1(\varphi_0 + \xi) + \dots - (\lambda_0 + \epsilon \sigma) N(\varphi_0 + \xi, \varphi_0 + \xi)\} = 0 \tag{4.2.7}$$

Equations (4.2.6) and (4.2.7) are equivalent to the original problem (4.2.5); that is, a solution of one gives a solution of the other. (4.2.6) and (4.2.7) are called the Lyapounov-Schmidt equations.

Now define

$$\mathcal{F}(\xi,\sigma,\epsilon) = \Big(\xi + \epsilon K\{ \}, \ \sigma[L_1 \varphi_0] + \sigma[L_1 \xi] + \dots - (\lambda_0 + \epsilon \sigma)[N(\varphi_0 + \xi, \varphi_0 + \xi)]\Big)$$

$\mathcal{F}$ is a mapping from $B \times R \times R$ to $B \times R$, where B is the Banach space $\{\xi : \xi \in C_{2+\alpha,\sigma}, [\xi] = 0\}$.
We leave it to the reader to verify that $\mathcal{F}(\xi,\sigma,\epsilon)$ is an analytic mapping of $\xi \in B$ and the scalar variables ϵ and σ. We wish to solve

$$\mathcal{F}(\xi,\sigma,\epsilon) \equiv 0$$

We have

$$\mathcal{F}(\xi,\sigma,0) = (\xi, \sigma[L_1\varphi_0] + \sigma[L_1\xi] - \lambda_0[N(\varphi_0 + \xi, \varphi_0 + \xi)]) \quad .$$

We want $\mathcal{F} = 0$ so take $\xi = 0$, and

$$\sigma_0[L_1\varphi_0] - \lambda_0[N(\varphi_0,\varphi_0)] = 0 \quad .$$

This determines $\sigma_0 = \sigma(0)$, since $[L_1\varphi_0] \neq 0$. To apply the implicit function theorem we must compute the Frechet derivative of $\mathcal{F}$ at $(0,\sigma_0,0)$. It is the matrix operator (see exercise 4, below)

$$\begin{pmatrix} I & 0 \\ \sigma_0[L_1 \cdot] - \lambda_0[M(\varphi_0, \cdot)] & [L_1\varphi_0] \end{pmatrix} \tag{4.2.8}$$

This operator is invertible provided $[L_1\varphi_0] \neq 0$, which is precisely condition (4.2.3). So we can apply the implicit function theorem Q.E.D.

The nature of the function $\sigma(\epsilon)$ determines the form of the bifurcation. Since $\tau = \epsilon\sigma$ we have non-trivial solutions of (4.2.2) for

$$\lambda(\epsilon) = \lambda_0 + \epsilon\,\sigma\,(\epsilon) \quad .$$

We get non-trivial solutions of (4.2.2) parametrized by $\epsilon : (\epsilon \varphi_0 + \epsilon \xi(\epsilon), \lambda_0 + \epsilon \sigma(\epsilon))$. There are solutions for $\lambda > \lambda_0$ if and only if $\epsilon \sigma(\epsilon) > 0$ for some values of ϵ near $\epsilon = 0$.

It remains to prove Lemma 4.2.1. This will be carried out in § 4.4.

Exercises:

1. Calculate $\mu'(\lambda_0)$ as in lemma 4.2.2 by perturbation series. That is, expand $L(\lambda) = L_0 + \tau L_1 + \ldots$, $\varphi(\lambda) = \varphi_0 + \tau\varphi_1 + \ldots$, and $\mu(\lambda) = \tau\mu_1 + \ldots$, and calculate the term μ_1 .

2. Carry out the details of showing that the Lyapounov-Schmidt equations and the original problem are equivalent.

3. Show that the operator $\mathcal{F}(\xi,\sigma,\epsilon)$ is analytic in ξ,σ , and ϵ on $B \times R \times R$ to $B \times R$. $(B = C_{\sigma,2+\alpha} \cap \{\xi : [\xi] = 0\})$

4. Verify the expression (4.2.8) for the Frechet derivative of $\mathcal{F}$. (Hint: to calculate the Frechet derivative of an analytic mapping $F(x)$ at x_0 , expand $F(x_0 + t v)$ in powers of t . The Frechet derivative of F at x_0 is simply the coefficient of t . In this case the Banach space is $B \times R$. The vector x_0 is $\xi = 0$, $\sigma = \sigma_0$. For $\underline{v}$ take $\underline{v} = (\eta,\varphi)$. The notation $\sigma_0[L_1 \cdot]$ in the Frechet derivative (4.2.8) denotes the linear functional $\eta \to \sigma_0[L_1\eta]$.)

5. Discuss the nature of the bifurcation curve $v(\epsilon)$, $\tau(\epsilon)$ in the three cases $\sigma_0 \neq 0$; $\sigma_0 = 0$, $\sigma_1 > 0$; $\sigma_0 = 0$, $\sigma_1 < 0$. $(\sigma(\epsilon) = \sigma_0 + \epsilon\sigma_1 + \ldots)$

6. Investigate the bifurcation of solutions of

$$Lu + \lambda u + f(\lambda,u) = 0$$

$$u\Big|_{\partial D} = g$$

where L is the second order elliptic operator discussed in Chapter II.

7. Prove a general bifurcation result for an equation of the form

$$L(\lambda)u + F(\lambda,u) = 0$$

under the following hypotheses:

(i) L, F are analytic in λ and u from a Banach space B_1 to B_2, where $B_1 \subset B_2$.

(ii) $L(\lambda)$ has an eigenfunction φ:

$$L(\lambda)\varphi(\lambda) = \mu(\lambda)\varphi(\lambda)$$

with $\varphi \in B_1$, $\mu(\lambda_0) = 0$, $\mu'(\lambda_0) \neq 0$.

(iii) The range of L_0 has codemension 1: there exists a $\varphi_0{}^*$ in $B_2{}^* \subset B_1{}^*$ such that $Lu = f$ is solvable if and only if $\langle f,\varphi_0{}^*\rangle = 0$. In that case there is a unique u in B_1 such that $\langle u,\varphi_0{}^*\rangle = 0$ and $Lu = f$. The mapping $f \to u$ is then bounded.

4.3 <u>Poincaré-Lindstedt series:</u>

Having proved the existence of an analytic one parameter family of solutions $z(\epsilon)$, $\tau(\epsilon)$ of (4.2.2) we can now determine the representation of those functions as power series:

$$z(\epsilon) = \epsilon\, z_1 + \epsilon^2\, z_2 + \dots$$

$$\tau(\epsilon) = \epsilon\, \tau_1 + \epsilon^2\, \tau_2 + \dots \tag{4.3.1}$$

Let us investigate the procedure for determining the coefficients. To simplify the calculations we assume that

$$L(\lambda) = L_0 + \tau L_1$$

and consider the simpler equation

$$(L_0 + \tau L_1)z + N(z,z) = 0 \tag{4.3.2}$$

Recall that $\epsilon = [z]$; therefore

$$\epsilon = \epsilon[z_1] + \epsilon^2[z_2] + \dots$$

which implies that $[z_1] = 1$, $[z_j] = 0$ for $j = 2,3, \dots$
Inserting the power series (4.3.1) in (4.3.2) we get

$$\sum_{n=1}^{\infty} \epsilon^n L_0 z_n + \sum_{n=2}^{\infty} \epsilon^n \sum_{k+j=n} \tau_k L_1 z_j + \sum_{n=2}^{\infty} \epsilon^n \sum_{k+j=n} N(z_j, z_k) = 0 .$$

Equating each coefficient of ϵ to zero we get successively

$$\begin{aligned} & L_0 z_1 = 0 && (0) \\ & L_0 z_2 + \tau_1 L_1 z_1 + N(z_1, z_1) = 0 && (1) \\ & \vdots \\ & L_0 z_n + \tau_{n-1} L_1 z_1 + \dots = 0 && (n) \end{aligned} \tag{4.3.3}$$

The dots in equation (4.3.3) (n) denote terms which depend only on $z_1, z_2, \ldots z_{n-1}$ and $\tau_1, \ldots \tau_{n-2}$. The first equation implies that $z_1 = c\varphi_0$, and since $[z_1] = 1$ we see that $c = 1$. Taking the bracket of the second equation in (4.3.3) gives

$$\tau_1[L_1\varphi_0] + [N(\varphi_0,\varphi_0)] = 0$$

Since $[L_1\varphi_0] \neq 0$, this equation determines τ_1 ; z_2 is then determined by the second equation. (The Fredholm alternative implies that $Lu = f$ is solvable if $[f] = 0$. Furthermore, there is a unique solution u such that $[u] = 0$. The normalization $[z_2] = 0$ determines z_2 uniquely.) At the nth stage, τ_{n-1} is determined by the condition that (recall $z_1 = \varphi_0$)

$$\tau_{n-1}[L_1\varphi_0] + [\ldots] = 0 \quad .$$

Then there is a unique solution z_n of (4.3.3) (n) such that $[z_n] = 0$. In this way, all the coefficients z_j and τ_j are uniquely determined. The convergence of the series is assured by theorem 4.2.3.

The procedure is the same in the case of more general analytic non-linearities. One applies the solvability condition for L_0 to determine the coefficients τ_j at each stage. The determination of the coefficients τ_j and z_j in this manner is similar to Lindstedt's procedure in Celestial Mechanics for the formal integration of a Hamiltonian system of equations of motion.

Exercise:

1. Let

$$A = \begin{pmatrix} 1 & 0 \\ 0 & 2 \end{pmatrix}, \qquad B = \begin{pmatrix} 1 & 1 \\ 0 & 1 \end{pmatrix}, \quad N(u,v) = (x_1x_2 + y_1y_2,\ 2x_1y_2) \ ,$$

where $u = (x_1,y_1)$, $v = (x_2,y_2)$. For what values of λ does the equation

$$(A - \lambda B)u + N(u,u) = 0$$

have bifurcation points at $u = 0$? Verify the Fredholm alternative for $L_0 = A - \lambda_0 B$ at one of the bifurcation points. Find $L_0{}^*$ and $\varphi_0{}^*$. (Use the inner product $(u,v) = x_1x_2 + y_1y_2$) .

Expand

$$u = \epsilon\, u_1 + \epsilon^2\, u_2 + \ldots$$

$$\lambda = \lambda_0 + \epsilon\, \lambda_1 + \epsilon^2\, \lambda_2 + \ldots$$

Find λ_1, λ_2 and u_1 . Find the recursion formulas for u_n and λ_n .

4.4 Fredholm Alternative for L_0.

In this section we sketch briefly the proof of lemma 4.2.1. The techniques used apply in general to elliptic operators on smoothly bounded domains. We consider L_0 on both the Hilbert space H_σ and the space $C_{\sigma,2+\alpha}$ of divergence free, Hölder vector fields.

(a) Adjoint. The formal adjoint to L_0 on H_σ is the operator $L_0{}^*$ obtained by integration by parts. For the inner product we take the complex inner product

$$(u,v) = \int_{\mathcal{D}} u_i \overline{v_i}\, dx \ .$$

Let

$$L_0 u = P_\sigma(\Delta u_i + \tilde{u}_j \frac{\partial u_i}{\partial x_j} + u_j \frac{\partial \tilde{u}_i}{\partial x_j}) \quad ,$$

where $\operatorname{div} \tilde{u} = 0$ and define

$$L_0{}^* v = P_\sigma(\Delta v_i - \tilde{u}_j \frac{\partial v_i}{\partial x_j} + v_j \frac{\partial \tilde{u}_j}{\partial x_i}) \quad .$$

Then $(L_0 u, v) = (u, L_0{}^* v)$ for all smooth vector fields u and v which vanish on $\partial\mathscr{D}$. (The reader should check this.)

(b) Resolvent Operator. An easy computation (exercise) shows that

$$\operatorname{Re}(L_0 u, u) = \operatorname{Re} \int_{\mathscr{D}} (\Delta u_i \bar{u}_i + \tilde{u}_j \frac{\partial u_i}{\partial x_j} \bar{u}_i + u_j \frac{\partial \tilde{u}_i}{\partial x_j} \bar{u}_i) dx$$

$$= - \|u\|^2 + \operatorname{Re} \int_{\mathscr{D}} \frac{\partial \tilde{u}_i}{\partial x_j} u_i \bar{u}_j \, dx$$

where

$$\|u\|^2 = \int_{\mathscr{D}} \frac{\partial u_i}{\partial x_j} \frac{\partial \bar{u}_i}{\partial x_j} dx \quad . \qquad \text{(sum on } i \text{ and } j \text{)}$$

Assuming $\left|\frac{\partial \tilde{u}_i}{\partial x_j}\right| \leq$ const. on $\mathscr{D}$, we have

$$\operatorname{Re}(-L_0 u, u) \geq \|u\|^2 - c|u|^2$$

($|u|^2 = (u,u)$). Therefore, for $\operatorname{Re} \lambda \geq c$, we have

$$\operatorname{Re}((\lambda - L_0)u, u) \geq \|u\|^2 \quad . \tag{4.4.1}$$

This inequality is sufficient to establish that for $\operatorname{Re} \lambda \geq c$, $R(\lambda, L_0) = (\lambda - L_0)^{-1}$ exists on H_σ and is compact (see Agmon, p. 99;

the proof is based on the Lax-Milgram theorem).

Furthermore, by lemma 4.1.1 applied to the operator $(L_0 - \lambda)$ for large $\operatorname{Re} \lambda$, we see that R_λ is also a bounded mapping from $C_{\sigma,\alpha}$ to $C_{\sigma,2+\alpha}$. (In fact, inequality (4.4.1) shows that $(L_0 - \lambda)\varphi = 0$ can have no nontrivial solutions.) Therefore R_λ is a compact mapping of $C_{\sigma,2+\alpha}$ to itself.

(c) The resolvent operator $R(\lambda,L_0)$ is analytic in λ except possibly at a discrete set of λ's, say $\lambda_1,\lambda_2, \ldots$ which do not cluster at any finite complex value. These singular values of λ are poles of $R(\lambda,L_0)$ and are the eigenvalues of L_0. The <u>resolvent</u> set of L_0 is the set of λ for which $R(\lambda,L_0)$ exists. It is an open set. The <u>spectrum</u> of L_0 is the complement of the resolvent set. In this case the spectrum of L consists only of eigenvalues. These facts follow from the compactness of the resolvent operator.

(d) On the Hilbert space H_σ, a general theorem of functional analysis (Riesz-Nagy, p. 304) says that

$$R(\lambda,L_0)^* = R(\bar{\lambda},L_0^*)$$

This relationship shows that λ is in the resolvent set of L_0 iff $\bar{\lambda}$ is in the resolvent set of L_0^*. Equivalently, λ is an eigenvalue of L_0 iff $\bar{\lambda}$ is an eigenvalue of L_0^*.

(e) <u>Fredholm Alternative</u>. On H_σ, let $L_0\varphi_0 = 0$. Then there exists in H_σ an adjoint eigenfunction $\varphi_0^* : L_0^*\varphi_0^* = 0$. Assume 0 is a simple eigenvalue of L_0. To solve $L_0u = f$, write the equation in the form

$$(\lambda - L_0)u = \lambda u - f ,$$

or

$$u - \lambda R_\lambda u = -R_\lambda f \tag{4.4.2}$$

Since R_λ is a bounded transformation from $C_{\sigma,\alpha}$ to $C_{\sigma,2+\alpha}$, $|R_\lambda f|_{2+\alpha} \leq c|f|_\alpha$. Now $\varphi_0 - \lambda R_\lambda \varphi_0 = 0$ so (4.4.2) is not invertible for arbitrary right hand side. Let $\mathfrak{B}$ denote the Banach space

$$\mathfrak{B} = \{u \in C_{\sigma,2+\alpha} ; \ [u] = 0\} \ .$$

The norm on $\mathfrak{B}$ is taken to be the Hölder norm $| \ |_{2+\alpha}$. If $u \in \mathfrak{B}$ then $[R_\lambda u] = (R_\lambda u, \varphi_0{*}) = (u, R_\lambda{*}\varphi_0{*}) = (u, \bar{\lambda}^{-1}\varphi_0{*}) = 0$. So $R_\lambda : \mathfrak{B} \to \mathfrak{B}$, and we can consider (4.4.2) as a functional equation in $\mathfrak{B}$. R_λ is still a compact operator on $\mathfrak{B}$, so we apply the Fredholm alternative. This time, however, $(I - \lambda R_\lambda)\varphi = 0$ has no nontrivial solutions in $\mathfrak{B}$ (the only solution is φ_0, which does not lie in $\mathfrak{B}$). So (4.4.2) is boundedly invertible on $\mathfrak{B}$ and we get a unique solution u in $\mathfrak{B}$ — that is, $[u] = 0$. (Provided, of course, that $[R_\lambda f] = 0$. But $[R_\lambda f] = (R_\lambda f, \varphi_0{*}) = (f, R_\lambda{*}\varphi_0{*}) = \lambda^{-1}(f, \varphi_0{*}) = 0$.) Denote by $\hat{K}$ the mapping $f \to u$ given by solving (4.4.2). Then $|\hat{K}f|_{2+\alpha} \leq c'|f|_\alpha$; and, since (4.4.2) is equivalent to solving $L_0 u = f$, we can write $\hat{K}L_0 u = \hat{K}f = u$, provided f and u belong to $\mathfrak{B}$.

Finally, define the projections P and Q by

$$Pu = [u]\varphi_0 \ , \quad Q = I - P \ .$$

Let $K = \hat{K}Q$. Then K is defined on the whole space $C_{\sigma,2+\alpha}$, and $KL_0 = \hat{K}QL_0 = \hat{K}L_0 Q = Q$. This completes the proof of Lemma 4.2.1.

Exercises

1. Let $Pu = [u]\varphi_0$ and $Q = I - P$. Show that P and Q commute with L_0. Show $(L_0 u,v) = (u,L_0^* v)$ for smooth u and v which vanish on $\partial\mathcal{D}$. Show P and Q are bounded in $C_{\sigma,2+\alpha}$.

2. Consider the operator L on the space of functions $u(\theta_1,\theta_2)$ 2π periodic in θ_1,θ_2:

$$Lu = \omega_1 \frac{\partial u}{\partial\theta_1} + \omega_2 \frac{\partial u}{\partial\theta_2}$$

where $\frac{\omega_1}{\omega_2}$ is irrational. What are the eigenfunctions and eigenvalues of L? Does L satisfy the Fredholm alternative? Discuss the solvability of the equation $Lu = f$. (See Moser; this operator illustrates the problem of small divisors in Celestial mechanics.)

3. Verify the calculation of (Lu,u) in (b).

** 4. Let L be an operator with the following property: There exists λ_0 such that $(L - \lambda_0)^{-1}$ is a compact operator on a Banach space B. Show $(\lambda - L)^{-1}$ is a meromorphic function of λ whose poles are of finite order and cannot cluster anywhere in the finite plane. The poles of $R_\lambda = (\lambda - L)^{-1}$ are the eigenvalues of L. The Fredholm alternative applies to the equation $(L - \lambda_0)u = f$ when λ_0 is an eigenvalue of L.
Hint: see Riesz-Nagy. Write

$$R(\lambda,L) = (I + (\lambda - \lambda_0)R_{\lambda_0})^{-1}R_{\lambda_0} = R_{\lambda_0}(I + (\lambda - \lambda_0)R_{\lambda_0})^{-1}$$

and apply the Fredholm theory to the operator $I + (\lambda - \lambda_0)R_{\lambda_0}$.

4.5 Formal Calculation of the Stability.

In this section we calculate the critical eigenvalue by perturbation methods. For simplicity, we consider the equation

$$(L_0 + \tau L_1)u + N(u,u) = 0 \quad . \tag{4.5.1}$$

The eigenvalue problem associated with the linearized stability problem for u is

$$(L_0 + \tau L_1)\varphi + M(u,\varphi) = \sigma\varphi \quad . \tag{4.5.2}$$

The eigenvalue problem (4.5.2) is arrived at in a purely formal manner. The derived operator $L(\lambda) + M(u,\cdot)$, where u is a solution of (4.5.1), is the analogue of the Jacobian $(\frac{\partial f_i}{\partial x_j})$ in the case of ordinary differential equations. (see theorem 1.1.2, Chapter I.) We will see how the linearized stability criterion can be justified in the case of the Navier-Stokes equations in Chapter VI.

When $\tau = 0$ and $u = 0$ we have $\sigma = 0$ and $\varphi = \varphi_0$ in (4.5.2) by our assumption that $\tau = 0$, $u = 0$ is a bifurcation point. The stability of the trivial solution $u = 0$ is determined by the eigenvalues of the operator $L_0 + \tau L_1$. We assume that the trivial solution is unstable for $\tau > 0$ and stable for $\tau < 0$. We wish to determine the critical eigenvalue $\sigma(\tau)$ of (4.5.2) for small τ, when u is a non-trivial solution of (4.5.1).

As we have seen, the solution branch of (4.5.1) can be expanded in a power series $(u = \epsilon w)$

$$w = \varphi_0 + \epsilon w_1 + \epsilon^2 w_2 + \ldots$$

$$\tau = \epsilon \tau_1 + \epsilon^2 \tau_2 + \ldots$$

For now, assume that σ and φ also have power series expansions

$$\sigma = \epsilon \sigma_1 + \epsilon^2 \sigma_2 + \dots$$
$$\varphi = \varphi_0 + \epsilon \varphi_1 + \dots \tag{4.5.3}$$

The convergence of these series is proved below. We normalize φ so that $[\varphi] \equiv 1$; this means that in the power series for φ , $[\varphi_j] = 0$ for $j = 1,2, \dots$.

<u>Theorem 4.5.1</u>: <u>If $\tau_1 \neq 0$ then</u>

$$\sigma_1 = - \tau_1[L_1\varphi_0] ; \tag{4.5.4}$$

<u>if $\tau_1 = 0$ then $\sigma_1 = 0$ and</u>

$$\sigma_2 = - 2 \tau_2[L_1\varphi_0] . \tag{4.5.5}$$

<u>Accordingly, subcritical branches are unstable and supercritical branches are stable if $\tau_1 \neq 0$ or if $\tau_1 = 0$ and $\tau_2 \neq 0$</u> . (We assume, of course, that all other eigenvalues of $L_0 + \tau L_1$ have negative real parts for small τ)

<u>Proof</u>: We have

$$(L_0 + \epsilon \tau_1 L_1 + \epsilon^2 \tau_2 L_1 + \dots)(\varphi_0 + \epsilon w_1 + \dots)$$
$$+ N(\varphi_0,\varphi_0) + \epsilon M(\varphi_0,w_1) + \dots = 0 ;$$

hence, equating the coefficient of each power of ϵ to zero, we get

$$L_0\varphi_0 = 0$$
$$\tau_1 L_1\varphi_0 + L_0 w_1 + N(\varphi_0,\varphi_0) = 0 \tag{4.5.6}$$

Taking brackets of the second equation gives (since $[L_0\psi] = 0$ for any ψ)

$$\tau_1[L_1\varphi_0] + [N(\varphi_0,\varphi_0)] = 0 \quad . \tag{4.5.7}$$

If $\tau_1 = 0$ then the third equation obtained is

$$\tau_2 L_1\varphi_0 + L_0 w_2 + M(\varphi_0, w_1) = 0 \quad .$$

Taking brackets of this equation, we get

$$\tau_2[L_1\varphi_0] + [M(\varphi_0,w_1)] = 0 \quad \text{if} \quad \tau_1 = 0 \ . \tag{4.5.8}$$

Now we calculate the perturbation terms for σ . By a similar procedure to the one above, we get the equation

$$L_0\varphi_1 + \tau_1 L_1\varphi_0 + M(\varphi_0,\varphi_0) = \sigma_1\varphi_0 \ ; \tag{4.5.9}$$

and, in case $\tau_1 = 0$, the next equation is

$$L_0\varphi_2 + \tau_2 L_1\varphi_0 + M(w_1,\varphi_0) + M(\varphi_0,\varphi_1) = \sigma_2\varphi_0 + \sigma_1\varphi_1 \ . \tag{4.5.10}$$

If $\tau_1 \neq 0$, then by taking brackets of (4.5.9) we obtain

$$\tau_1[L_1\varphi_0] + [M(\varphi_0,\varphi_0)] = \sigma_1 \quad . \tag{4.5.11}$$

From (4.5.7) and the fact that $M(u,u) = 2N(u,u)$ we get equation (4.5.4).

If $\tau_1 = 0$ then (4.5.7) and (4.5.11) show that $\sigma_1 = 0$. Taking brackets of (4.5.10), we then see that

$$\tau_2[L_1\varphi_0] + [M(w_1,\varphi_0)] + [M(\varphi_0,\varphi_1)] = \sigma_2 \quad .$$

From this and (4.5.8) we then see that $\sigma_2 = [M(\varphi_0,\varphi_1)]$. Since

$\tau_1 = \sigma_1 = 0$, φ_1 satisfies (see (4.5.9))

$$L_0\varphi_1 + 2N(\varphi_0,\varphi_0) = 0 ;$$

while w_1 satisfies (4.5.6) with $\tau_1 = 0$. Since $[\varphi_1] = [w_1] = 0$, we see that $\varphi_1 = 2w_1$. Hence

$$\sigma_2 = 2[M(\varphi_0,w_1)] = - 2\tau_2[L_1\varphi_0]$$

(see (4.5.8)).

The final statement of the theorem follows from the fact that $[L_1\varphi_0] > 0$ if the null solution loses stability for $\tau > 0$. The reader can now easily verify the exchange of stability diagrams on p. 5.

It remains to prove the convergence of the series (4.5.3). We do this by use of an implicit function theorem argument similar to that used in the bifurcation scheme. We write $\varphi = \varphi_0 + \xi$ where $[\xi] = 0$ and substitute in (4.5.2) to get

$$L_0\xi + \tau L_1\varphi_0 + \tau L_1\xi + M(u,\varphi_0 + \xi) = \sigma\varphi_0 + \sigma\xi \quad .$$

Operating on this equation by K we get

$$\xi + \tau KL_1\varphi_0 + \tau KL_1\xi + KM(u,\varphi_0 + \xi) - \sigma K\varphi_0 - \sigma K\xi = 0 \qquad (4.5.12)$$

and taking the bracket we get

$$- \sigma + \tau[L_1\varphi_0] + \tau[L_1\xi] + [M(u,\varphi_0 + \xi)] = 0 \qquad (4.5.13)$$

We now proceed as before. The equations (12) and (13) for ξ and σ constitute a functional equation $\mathcal{F}(\xi,\sigma,\epsilon) = 0$. Here $\mathcal{F} : B \times R \times R$ into $B \times R$ and consists of the left sides of (4.5.12) and (4.5.13).

(Recall $\tau = \tau(\epsilon)$ and $u = u(\epsilon)$) . Again, σ is a real number and ξ lies in the subspace $B = \{\xi : \xi \in C_{\sigma,2+\alpha}, [\xi] = 0\}$. For $\epsilon = 0$ we have $\mathcal{F}(0,0,0) = 0$, while the Frechet derivative of $\mathcal{F}$ at $\epsilon = 0$ is (Note that $\tau = u = 0$ when $\epsilon = 0$.)

$$\begin{pmatrix} I & -K\varphi_0 \\ 0 & -1 \end{pmatrix} \tag{4.5.14}$$

This operator is invertible on the Banach space $\mathcal{B} \times R$. So the implicit function theorem applies and we get $\xi = \xi(\epsilon)$ and $\sigma = \sigma(\epsilon)$. Since the operations in (4.5.12) and (4.5.13) are analytic, so are ξ and σ .

<u>Exercises</u>:

1. Verify (4.5.14) for the Frechet derivative, and show that all the operations in (4.5.12) are analytic.

2. Carry out the formal calculations for the first nonvanishing terms for the critical eigenvalue in the case of the more general functional equation

$$L(\lambda)u = F(\lambda,u)$$

where F is analytic in λ and u ; and $F(\lambda,0) \equiv 0$, $F_u(\lambda,0) \equiv 0$.

Notes

The assumption that L_0 has a simple eigenvalue in general cannot be relaxed. Topological degree arguments show that if the multiplicity of the zero eigenvalue is odd then bifurcation must necessarily occur; but it is possible that no bifurcation occurs at eigenvalues of even multiplicity. The theory of bifurcation at multiple eigenvalues is much more complicated and the picture is not so simple or complete as it is in the case of simple eigenvalues. See the articles by D. Sather[1] and K. Kirchgässner[2]. In Chapter VII, using a topological degree argument we shall prove that bifurcation takes place at a simple eigenvalue without the assumption $\mu'(\lambda_0) \neq 0$ (4.2.3). This assumption, however, does guarantee that the bifurcation consists of a simple curve of non-trivial solutions. In the general case we may get several solution curves intersecting at the bifurcation point.

Examples of bifurcation of stationary solutions in hydrodynamics problems will be discussed in Chapter VIII.

Exercise 2 on p.95 gives a simplified illustration of a fundamental problem in Celestial Mechanics known classically as "the problem of small divisors". It amounts to a failure of the Fredholm alternative. The eigenvalues of L are the numbers $\{i(m\omega_1 + n\omega_2)\}$ where m and n are integers and ω_2/ω_1 is irrational. 0 is a simple eigenvalue of L but the non-zero eigenvalues of L are dense on the imaginary axis. The formal solvability condition $[f] = 0$ amounts to

$$[f] = \int_0^{2\pi} \int_0^{2\pi} f(\theta_1,\theta_2)\, d\theta_1 d\theta_2 = 0 \, .$$

If $[f] = 0$ we can formally solve $Lu = f$ by Fourier series. But the mapping $f \to u$ will not be a bounded transformation (say on $L_2((0,2\pi) \times (0,2\pi))$. See the article by J. Moser.

M. Reeken [1,2] has recently proved that for gradient operators with an eigenvalue of multiplicity m there are at least $2m$ solutions branching from the point of bifurcation.

V

Bifurcation of Periodic Solutions

5.1 Introduction:

In this chapter we treat the transition to instability when an equilibrium solution loses stability by virtue of a complex conjugate pair of simple eigenvalues crossing the imaginary axis. We again treat the Navier-Stokes equations in operator form

$$\frac{\partial u}{\partial t} + L(\lambda)u + N(u,u) = 0 \quad . \tag{5.1.1}$$

We assume here, for simplicity, that $L(\lambda) = L_0 + \tau L_1$, where $\tau = \lambda - \lambda_0$ and L_0 is the operator discussed at length in Chapter IV (see § 4.4).

The null solution $u = 0$ is an equilibrium (time independent) solution of (5.1.1); according to the principle of linearized stability, its stability is determined by the spectrum of the operator $L(\lambda)$. We assume that $L(\lambda)$ has a simple eigenvalue $\gamma(\lambda)$ such that $\gamma(\lambda_0) = i$ and $\operatorname{Re} \gamma'(\lambda_0) < 0$. Under these conditions we wish to investigate the bifurcation of periodic solutions of (5.1.1). In constructing the periodic solutions the period, as well as the amplitude of the oscillations, must be determined in terms of $\lambda - \lambda_0$. This is accomplished in the following way: put $s = \omega t$ and $u = \epsilon v$ in (5.1.1); then that equation becomes

$$\omega \frac{\partial v}{\partial s} + L(\lambda)v + \epsilon N(v,v) = 0 \tag{5.1.2}$$

We are going to construct a one parameter family $v = v(s,\epsilon)$, $\omega = \omega(\epsilon)$,

$\lambda = \lambda(\epsilon)$ of solutions of (5.1.2), where v is 2π periodic in s. This gives a family of $2\pi/\omega$ periodic solutions of (5.1.1), namely $u(t,\epsilon) = \epsilon\, v(\omega(\epsilon)t,\epsilon)$. In this way, ω, λ, and u are determined as functions of ϵ, and indirectly, ω and u are given as functions of $\lambda - \lambda_0$.

The procedure for constructing periodic solutions bears some similarities to that for constructing bifurcating stationary solutions, as in Chapter IV.

5.2 Riesz-Schauder theory for $\frac{\partial}{\partial s} + L_0$.

Let $\xi(\lambda)$ and $\gamma(\lambda)$ denote the critical eigenfunction and eigenvalue of $L(\lambda)$: thus, $L(\lambda)\xi(\lambda) = \gamma(\lambda)\xi(\lambda)$, and $\gamma(\lambda_0) = i$, $\operatorname{Re} \gamma'(\lambda_0) < 0$. The operator $L(\lambda)$ is easily seen to commute with complex conjugation: $\overline{L(\lambda)u} = L(\lambda)\bar{u}$. Consequently the eigenvalues of $L(\lambda)$ appear in complex conjugate pairs. Letting $\xi_0 = \xi(\lambda_0)$ we have

$$L_0\xi_0 = i\,\xi_0 \quad \text{and} \quad L_0\bar{\xi}_0 = -i\,\bar{\xi}_0 .$$

From the Riesz-Schauder theory for L_0 (Lemma 4.2.1) there exists a vector field $\xi_0{}^*$ such that

$$L_0\xi_0{}^* = -i\,\xi_0{}^* \quad \text{and} \quad L_0{}^*\overline{\xi_0{}^*} = i\,\overline{\xi_0{}^*} .$$

Since i is a simple eigenvalue we may normalize $\xi_0{}^*$ so that $(\xi_0, \xi_0{}^*) = 1$. The notation $(\,,\,)$ denotes, as usual, the inner product on H_σ.

Let us compute $\operatorname{Re} \gamma'(\lambda_0)$. We write

$$\xi(\lambda) = \xi_0 + \tau\,\xi_1 + \dots$$

$$\gamma(\lambda) = i + \tau\,\gamma_1 + \dots$$

Substituting these series into $L(\lambda)\xi(\lambda) = \gamma(\lambda)\xi(\lambda)$ we obtain

$$L_0\xi_0 = i\,\xi_0$$

and

$$L_0\xi_1 + L_1\xi_0 = i\,\xi_1 + \gamma_1\xi_0 \quad .$$

We now take the inner product of the second equation with $\xi_0{}^*$. Since $(L_0\xi_1, \xi_0{}^*) = (\xi_1, L_0{}^*\xi_0{}^*) = (\xi_1, -i\,\xi_0{}^*) = i(\xi_1, \xi_0{}^*)$ and since $(\xi_0, \xi_0{}^*) = 1$ we obtain

$$\operatorname{Re}\gamma_1 = \operatorname{Re}\gamma'(\lambda_0) = \operatorname{Re}(L_1\xi_0, \xi_0{}^*) \neq 0 \qquad (5.2.1)$$

We introduce the following Banach spaces:

(1) $\mathcal{P}_{2\pi}$ is the Hilbert space of (complex valued) 2π-periodic functions with values in H_σ; the inner product on $\mathcal{P}_{2\pi}$ is

$$(u,v)_{2\pi} = \frac{1}{2\pi}\int_0^{2\pi} (u(s), v(s))ds \quad ,$$

where

$$(u(s), v(s)) = \int_D u_i(x,s)\,\overline{v_i(x,s)}\,d\underline{x} \quad .$$

(2) $\mathcal{P}_{2\pi}^{k+2\alpha,\, \ell+\alpha}$ is the Banach space of <u>real</u> 2π periodic vector fields with finite Hölder norms $|\ |_{k+2\alpha,\, \ell+\alpha}$ $(k = 0,1,2,\ \ \ell = 0,1.)$ The first subscript refers to space derivatives, the second to time derivatives. Thus, the norm $|\ |_{2\alpha,\alpha}$ $(k = 0,\ \ \ell = 0)$ is given by

$$|u|_{2\alpha,\alpha} = \sup_{\substack{x \in D \\ 0 \le s \le 2\pi}} \sum_{i=1}^{3} |u_i(x,s)| + \sup_{\substack{x,y \in D \\ 0 \le s,t \le 2\pi}} \sum_{i=1}^{3} \frac{|u_i(x,s) - u_i(y,t)|}{|x-y|^{2\alpha} + |s-t|^{\alpha}} .$$

The other norms are computed by applying the norm $|\ |_{2\alpha,\alpha}$ to the function u and its space derivatives up to order k and time derivatives to order ℓ, and summing.

Let $J = \frac{\partial}{\partial s} + L_0$. The homogeneous equation $J\varphi = 0$ has the nontrivial solutions $z_1(s) = e^{-is}\xi_0$ and $z_2 = \bar{z}_1$. (real solutions Re z_1 and Im z_1.) The adjoint operator $J^* = -\frac{\partial}{\partial s} + L_0^*$ has the null functions $z_1^* = e^{-is}\xi_0^*$ and $z_2^* = \overline{z_1^*}$. We wish to establish the Riesz-Schauder theory for J on the Banach space $P_{2\pi}^{2+2\alpha,\ 1+\alpha}$.

Introducing the notation $[u] = (u, z_1^*)_{2\pi}$, we prove the following theorem.

<u>Theorem 5.2.1</u>: <u>The equation $Ju = f$ is solvable for f in $P_{2\pi}^{2\alpha,\alpha}$ iff $[f] = 0$. If $[f] = 0$ then there is a unique solution u in $P_{2\pi}^{2+2\alpha,\ 1+\alpha}$ such that $[u] = 0$. If the transformation $f \to u$ be denoted by K, then $|Kf|_{2+2\alpha,\ 1+\alpha} \le c|f|_{2\alpha}$ for some constant c independent of f. There is a projection P onto the null space of J and $Q = I - P$ is a projection onto the range of J.</u>

<u>Remark</u>: The null space of J is two dimensional, but if f is real ($P_{2\pi}^{2\alpha,\alpha}$ is a real Banach space), then $\overline{[f]} = (\overline{f, z_1^*})_{2\pi} = (f, \overline{z_1^*})_{2\pi} = (f, z_2^*)_{2\pi}$; so $[f] = 0$ contains two orthogonality conditions. We break down the proof of Theorem 5.2.1 into a series of lemmas.

Lemma 5.2.2: (Solonnikov). Let u_i, p satisfy the initial value problem

$$\frac{\partial u_i}{\partial t} - \frac{1}{R}\,\Delta u_i = -\frac{\partial p}{\partial x_i} + f_i \tag{5.2.2}$$

$$\frac{\partial u_i}{\partial x_i} = 0\ , \quad u_i(x,0) = \psi_i(x)\ , \quad u_i\Big|_{\partial D} = 0\ .$$

Let $\operatorname{div}\psi = 0$, $\psi \in C_{2+2\alpha}(\bar{D})$, $f \in C_{2\alpha,\alpha}(\bar{D}_T)$, $(\bar{D}_T = D \times [0,T])$, and let $\partial D \in C_{2+2\alpha}$. Then $u \in C_{2+2\alpha,\ 1+\alpha}(\bar{D}_T)$ and there is a constant c such that $(|u|_{2+2\alpha,\ 1+\alpha} + |p|_{1+2\alpha,\ 1+\alpha}) \leq c(|f|_{2\alpha,\alpha} + |\psi|_{2+2\alpha})$.

The proof of this lemma, which is quite technical, requires the use of hydrodynamic potentials. (See Ladyzhenskaya, p. 100 for a discussion and references).

Lemma 5.2.3: Let $f \in P_{2\pi}^{2\alpha,\alpha}$. Then there exists a 2π periodic vector field $u \in P_{2\pi}^{2+2\alpha,\ 1+\alpha}$ which satisfies

$$\frac{\partial u_i}{\partial s} - \frac{1}{R}\,\Delta u_i = -\frac{\partial p}{\partial x_i} + f_i \tag{5.2.3}$$

$$\frac{\partial u_i}{\partial x_i} = 0 \qquad u_i\Big|_{\partial D} = 0\ .$$

Furthermore, there is a constant c independent of f , such that $|u|_{2+2\alpha,\ 1+\alpha} \leq c|f|_{2\alpha,\alpha}$.

Proof: Operating on (5.2.3) with the projection P_σ we may write it in the form

$$\frac{\partial u}{\partial s} + Au = f \tag{5.2.4}$$

We denote the semi-qroup generated by A by e^{-sA}. That is, the solution of

$$\frac{\partial u}{\partial s} + Au = 0 \quad u(0) = u_0$$

at time s is denoted by $e^{-sA}u_0$. In terms of the semi-group e^{-sA}, the solution of (5.2.4) with $u(0) = 0$ has the representation

$$u(s) = \int_0^s e^{-(s-t)A} f(t)dt \ .$$

In the appendix (§ 5.5) we shall show that

$$|e^{-sA}u_0|_{2+2\alpha} \leq c_1|u_0|_{2\alpha} \tag{5.2.5}$$

where c_1 depends on s. From lemma 5.2.2 we have

$$|\int_0^s e^{-(s-t)A} f(t)dt|_{2+2\alpha,\ 1+\alpha} \leq c_2|f|_{2\alpha,\alpha} \ . \tag{5.2.6}$$

The general solution of (5.2.4) can be represented in the form

$$u(s) = e^{-sA}\ u(0) + \int_0^s e^{-(s-t)A} f(t)dt \ , \tag{5.2.7}$$

and the condition that u be 2π periodic leads to

$$(I - e^{-2\pi A})u(0) = \int_0^{2\pi} e^{-(2\pi - t)A} f(t)dt \tag{5.2.8}$$

The estimate (5.2.5) for $s = 2\pi$ shows that the operator $e^{-2\pi A}$ is compact on $C_{2+2\alpha,\sigma}(D)$. Furthermore, $(I - e^{-2\pi A})\varphi = 0$ can have no nontrivial solutions φ in $C_{2+2\alpha,\sigma}$. (proof left as an exercise). Therefore (5.2.8) is invertible, and the solution u_0 satisfies $|u_0|_{2+2\alpha} \leq c|f|_{2\alpha,\alpha}$, (as a consequence of 5.2.6). With this bound on u_0 we get the required estimate on u from Lemma 5.2.2.

We now investigate the equation

$$(J + \lambda)u = f \qquad (5.2.9)$$

on $\mathcal{P}_{2\pi}^{2+2\alpha,\ 1+\alpha}$. We wish to show that (5.2.9) is invertible for large positive λ. Writing $Bu = \frac{\partial u}{\partial s} + Au$ we have $Ju = Bu + M(\tilde{u},u)$. Equation (5.2.9) can therefore be written

$$u + B^{-1}M(\tilde{u},u) + \lambda B^{-1}u = B^{-1}f \qquad (5.2.10)$$

Lemma 5.2.3 shows that B^{-1} exists and is a bounded operator from $C_{2\alpha,\alpha}$ to $C_{2+2\alpha,\ 1+\alpha}$. Accordingly, by the Fredholm alternative, it is sufficient to show that the equation $(J + \lambda)\varphi = 0$ has no non-trivial solutions. (Why?) However,

$$((J + \lambda)\varphi,\ \varphi)_{2\pi} = \frac{1}{2\pi}\int_0^{2\pi}\{(\frac{\partial\varphi}{\partial s},\ \varphi) + (A\varphi,\ \varphi) + (M(\tilde{u},\varphi),\varphi) + \lambda(\varphi,\varphi)ds$$

$$= \frac{1}{2\pi}\int_0^{2\pi}\{\frac{\partial}{\partial s}\frac{|\varphi|^2}{2} + \frac{1}{R}\,||\varphi||^2 + \int_D \varphi_i\varphi_j\,\frac{\partial\tilde{u}_i}{\partial x_j}\,d\underline{x}$$

$$+ \lambda\int_D \varphi_i\varphi_i\ d\underline{x}\}\ ds \quad ,$$

where

$$||\varphi||^2 = (A\varphi,\varphi) = \int_D \frac{\partial\varphi_i}{\partial x_j}\frac{\partial\varphi_i}{\partial x_j}\,d\underline{x} \quad .$$

(Since $\mathcal{P}_{2\pi}^{2+2\alpha,\ 1+\alpha}$ is a real Banach space, we can restrict ourselves to the invertibility of (5.2.9) on a real Banach space; hence φ may be taken to be real).

Since the derivatives $\frac{\partial\tilde{u}_i}{\partial x_j}$ are uniformly bounded, we have

$$((J + \lambda)\varphi,\varphi) \geq \frac{1}{2\pi}\int_0^{2\pi} \frac{1}{R} \|\varphi\|^2 + \lambda|\varphi|^2 - c|\varphi|^2 d\underline{x} \quad .$$

for some constant c . Thus, for $\lambda > c$, $(J + \lambda)\varphi = 0$ has no non-trivial solutions.

We are now in a position to prove theorem 5.2.1. Denote by R_λ the transformation $(J + \lambda)^{-1}$. The previous argument shows that for large positive λ , R_λ exists and is a bounded transformation from $P_{2\pi}^{2\alpha,\alpha}$ to $P_{2\pi}^{2+2\alpha,\ 1+\alpha}$. (In fact, in (5.2.10), $|B^{-1}f|_{2+2\alpha,\ 1+\alpha} \leq c|f|_{2\alpha,\alpha}$, and the left side is boundedly invertible on $P_{2\pi}^{2+2\alpha,\ 1+\alpha}$.)

We now write the equation $Ju = f$ in the form $(J + \lambda)u = f + \lambda u$, which is equivalent to

$$u - \lambda R_\lambda u = R_\lambda f \quad . \tag{5.2.11}$$

Now if $[f] = 0$, then $[R_\lambda f] = (R_\lambda f,\ z_1{}^*)_{2\pi} = (f,R_\lambda{}^* z_1{}^*) = (f,\frac{1}{\lambda}\ z_1{}^*) = 0$. Furthermore, $|R_\lambda f|_{2+2\alpha,\ 1+\alpha} \leq c|f|_{2\alpha,\alpha}$. So $R_\lambda f$ lies in the closed subspace

$$\mathcal{M}_{2\pi} = \{w : w \in P_{2\pi}^{2+2\alpha,\ 1+\alpha} \ ,\ [w] = 0\} \quad .$$

We now consider (5.2.11) on the Banach space $\mathcal{M}_{2\pi}$. R_λ is compact so we may apply the Fredholm alternative. But the only solutions of $\varphi - \lambda R_\lambda \varphi = 0$ are $\varphi = \operatorname{Re} z_1$ and $\varphi = \operatorname{Im} z_1$ which do not lie in $\mathcal{M}_{2\pi}$. In fact, $[z_1] = 1$, $[z_2] = 0$, so $[\operatorname{Re} z_1] = [z_1 + z_2] = 1$. Therefore (5.2.11) is boundedly invertible on $\mathcal{M}_{2\pi}$, and there is a constant c such that $|u|_{2+2\alpha,\ 1+\alpha} \leq c|R_\lambda f|_{2+2\alpha,\ 1+\alpha} \leq c'|f|_{2\alpha,\alpha}$. That the solution u is unique follows easily; for if $Ju = Ju' = f$ then $Jw = 0$ and $[w] = 0$, where $w = u - u'$. But $Jw = 0$ means that $w = az_1 + \bar{a}z_2$, (since w is real); and $[w] = a[z_1] + \bar{a}[z_2] = a = 0$, so $w = 0$.

We denote by the K transformation $f \to u$ defined by solving $Ju = f$. Then K is defined on the Banach space $\{f : [f] = 0 , f \in P_{2\pi}^{2\alpha,\alpha}\}$ and maps this space into $\mathcal{M}_{2\pi}$. The projection $Pu = 2 \operatorname{Re} [u]z_1$ is the required projection onto the null space of J . (Proof left as an exercise.)

<u>Exercises</u>:

1. Show that the equation $(I - e^{-2\pi A})\varphi = 0$ can have no nontrivial solutions. Hint: $\varphi = e^{-2\pi A}\varphi$ is equivalent to the existence of a 2π periodic solution of

$$\frac{\partial u_i}{\partial t} - \frac{1}{R}\Delta u_i = -\frac{\partial p}{\partial x_i}$$

$$\frac{\partial u_i}{\partial x_i} = 0 \qquad u_i\Big|_{\partial D} = 0 .$$

2. Compute $(z_i, z_j^*)_{2\pi}$ and $[z_i]$. Show that $Pu = 2 \operatorname{Re} [u]z_1$ is a projection onto the null space of J .

3. Prove that, for real λ , $L(\lambda)$ commutes with the operation of complex conjugation.

4. Prove that $R_\lambda z_1 = \frac{1}{\lambda} z_1$ and $R_\lambda^* z_1^* = \frac{1}{\bar{\lambda}} z_1^*$.

5.3. Solution of the Bifurcation problem:

We now construct the one parameter family ω , λ , v of periodic solutions of (5.1.2), with v lying in the Banach space $P^{2+2\alpha,\ 1+\alpha}_{2\pi}$. First, note we can assume that $[v] > 0$. In fact, letting $(T_\delta v)(s,\epsilon) = v(s+\delta,\epsilon)$ we see that $[T_\delta v] = e^{-i\delta}[v]$, so by an appropriate phase shift we can assume $[v] > 0$. Furthermore, we can assume $[v] = 1$ (this in effect normalizes our choice of the parameter ϵ).

For further reference we note the following relationships, both easily deriveable:

$$\text{(i)}\quad [z_1] = 1\ ,\quad [z_2] = 0$$

$$\text{(ii)}\quad [\tfrac{\partial w}{\partial s}] = -i[w] \quad \text{for any } 2\pi \text{ periodic vector field } w \tag{5.3.1}$$

We write equation (5.1.2) in the form

$$Jv + (\omega - 1)\frac{\partial v}{\partial s} + \tau L_1 v + \epsilon N(v,v) = 0 \tag{5.3.2}$$

Letting $v = Pv + Qv = Pv + \varphi$ we have $[\varphi] = 0$, since φ is in the range of J , and $Pv = 2\mathrm{Re}\,[v]z_1 = 2\mathrm{Re}\ z_1$, since $[v] = 1$. We denote $\mathrm{Re}\ z_1$ by u_0 and write $v = u_0 + \varphi$. Substituting this into (5.3.2) we get

$$J\varphi + (\omega - 1)\frac{\partial u_0}{\partial s} + (\omega - 1)\frac{\partial\varphi}{\partial s} + \tau L_1 u_0 + \tau L_1\varphi + \epsilon N(u_0 + \varphi,\ u_0 + \varphi) = 0$$

We now apply the projections P and Q to this equation to obtain (see the remarks below)

$$J\varphi + (\omega - 1)\frac{\partial\varphi}{\partial s} + \tau\ QL_1 u_0 + \tau\ QL_1\varphi + \epsilon\ QN(u_0 + \varphi,\ u_0 + \varphi) = 0 \tag{5.3.4}$$

$$(\omega - 1)[\frac{\partial u_0}{\partial s}] + (\omega - 1)[\frac{\partial \varphi}{\partial s}] + \tau[L_1 u_0] + \tau[L_1\varphi] + \epsilon[N(u_0 + \varphi, u_0 + \varphi)] = 0 \qquad (5.3.5)$$

By using (5.3.1), (ii) we can put the second equation in the form

$$-i(\omega - 1) + \tau[L_1 u_0] + \tau[L_1\varphi] + \epsilon[N(u_0 + \varphi, u_0 + \varphi)] = 0.$$

Taking real and imaginary parts of this equation we obtain

$$\tau \operatorname{Re}[L_1 u_0] + \tau \operatorname{Re}[L_1\varphi] + \epsilon \operatorname{Re}[N(u_0 + \varphi, u_0 + \varphi)] = 0 \qquad (5.3.6)$$

$$-(\omega - 1) + \tau \operatorname{Im}[L_1 u_0] + \tau \operatorname{Im}[L_1\varphi] + \epsilon \operatorname{Im}[N(u_0 + \varphi, u_0 + \varphi)] = 0 . \qquad (5.3.7)$$

Remarks: 1. In obtaining (5.3.4) we have used the facts that: $QJw = Jw$ for any w ; $Q \frac{\partial \varphi}{\partial s} = \frac{\partial \varphi}{\partial s}$ if φ is in the range of J ; and $Q \frac{\partial u_0}{\partial s} = 0$ since u_0 is in the null space of J .

2. In obtaining (5.3.5) we have used the fact that $Pw = 0$ implies $[w] = 0$.

Now we apply the transformation K given by Theorem 5.2.1 to equation (5.3.4) to obtain

$$\varphi + (\omega - 1)K \frac{\partial \varphi}{\partial s} + \tau\, K\, QL_1 u_0 + \tau\, K\, QL_1\varphi + \epsilon\, KQN(u_0 + \varphi, u_0 + \varphi) = 0 \qquad (5.3.8)$$

Since K picks up two space derivatives and one time derivative, each of the operations in (5.3.8) is bounded. (The reader can easily check that P , hence Q , is a bounded transformation from $\wp_{2\pi}^{k+2\alpha,\, \ell+\alpha}$ to itself).

The equations (5.3.6) - (5.3.8) are the equivalent of the Lyapounov-Schmidt equations of Chapter IV. We define a mapping $\mathcal{F}(\phi,\omega,\tau,\epsilon) = (\mathcal{F}_1,\mathcal{F}_2,\mathcal{F}_3)$ with $\mathcal{F}_1$ the left side of (5.3.8), $\mathcal{F}_2$ the left side of (5.3.7) and $\mathcal{F}_3$

the left side of (5.3.6). Let B be the Banach space $\{\varphi : \varphi \in \rho_{2\pi}^{2+2\alpha,\ 1+\alpha}, \ [\varphi] = 0\}$. Then $\mathcal{F}$ is a mapping from $B \times R \times R \times R$ to $B \times R \times R$. The reader can easily verify that $\mathcal{F}$ is analytic. We have $\mathcal{F}(0, 1, 0, 0) = 0$ while the Frechet derivative of $\mathcal{F}$ at $(0, 1, 0, 0)$ is

$$\begin{bmatrix} \frac{\partial \mathcal{F}_1}{\partial \phi} & \frac{\partial \mathcal{F}_1}{\partial \tau} & \frac{\partial \mathcal{F}_1}{\partial \omega} \\ \frac{\partial \mathcal{F}_2}{\partial \phi} & \frac{\partial \mathcal{F}_2}{\partial \tau} & \frac{\partial \mathcal{F}_2}{\partial \omega} \\ \frac{\partial \mathcal{F}_3}{\partial \phi} & \frac{\partial \mathcal{F}_3}{\partial \tau} & \frac{\partial \mathcal{F}_3}{\partial \omega} \end{bmatrix} = \begin{bmatrix} I & KQL_1u_0 & 0 \\ 0 & \mathrm{Re}[L_1u_0] & 0 \\ 0 & \mathrm{Im}[L_1u_0] & -1 \end{bmatrix} . \quad (5.3.9)$$

This operator is invertible provided $\mathrm{Re}[L_1u_0] \neq 0$; however, a simple computation shows that $\mathrm{Re}[L_1u_0] = \mathrm{Re}\ \gamma'(\lambda_0) = \mathrm{Re}(L_1\xi_0,\xi_0^*)$, which is non-zero by assumption (see 5.2.1).

This completes the construction of the bifurcating family of periodic solutions of (5.1.2).

Exercises:

1. Verify that $\mathrm{Re}[L_1u_0] = \mathrm{Re}(L_1\xi_0,\xi_0^*)$, where $u_0 = \mathrm{Re}\ e^{-is}\xi_0$.

2. Check that the operations in (5.3.8) are in fact continuous. (including the nonlinear term).

3. Check the computation of the Frechet derivative (5.3.9).

5.4 Stability; Floquet exponents.

In the case of ordinary differential equations the stability of a periodic solution of an autonomous system

$$\dot{x} = F(x) \tag{5.4.1}$$

can be determined by computing the Floquet exponents of the variation equations

$$\dot{y} = F'(x(t))y \quad \text{or} \quad \dot{y}_i = \frac{\partial F_i}{\partial x_j}(x(t))y_j \ , \tag{5.4.2}$$

where $x(t)$ is the periodic solution in question, and $F'(x(t))$ denotes the Frechet derivative (Jacobian) of F at $x(t)$.

The Floquet exponents are determined as follows: Suppose we are given a linear system of ordinary differential equations with periodic coefficients

$$\dot{y} = A(t)y \ , \quad \text{where} \quad A(t) = A(t+T) \ . \tag{5.4.3}$$

Denote by $\Phi(t)$ the fundamental solution matrix; thus

$$\dot{\Phi}(t) = A(t)\Phi \quad \text{and} \quad \Phi(0) = I \ .$$

The eigenvalues of $\Phi(T)$ are called the Floquet multipliers of (5.4.3). If μ is a multiplier, then there is a vector ψ such that $\Phi(T)\psi = \mu\psi$. Equivalently, if $\dot{z}(t) = A(t)z$ and $z(0) = \psi$, then $z(T) = \mu\psi = \mu z(0)$. Put $\mu = e^{-\beta T}$ and $w(t) = e^{\beta t}z(t)$. Then $w(T) = e^{\beta T}z(T) = \mu^{-1}z(T) = z(0) = w(0)$, and

$$\dot{w} - \beta w - A(t)w = 0 \ .$$

Thus the <u>Floquet exponent</u> β can be considered an eigenvalue of the problem

$$\dot{w} - A(t) - \beta w = 0$$
$$w(0) = w(T) \tag{5.4.4}$$

If $\text{Re}\,\beta > 0$ then $|\mu| < 1$ and $|z(T)| < |z(0)|$. If all the Floquet exponents have positive real parts, then all the eigenvalues of $\Phi(T)$ are less than one in absolute value, and $\Phi(T)$ is a contraction mapping. In the case of a periodic solution $x(t)$ of (5.4.1), however, $y(t) = \dot{x}(t)$ is a periodic solution of the variation equations; hence one of the Floquet multipliers is alway unity. Nevertheless, <u>if all the remaining Floquet multipliers of the variational equation (5.4.2) are less than one in absolute value the periodic solution $x(t)$ is orbitally stable</u>. (See Coddington and Levinson)

To be precise, let $y(t,y_0)$ denote the solution of $\dot{y} = F(y)$, $y(0) = y_0$. Then $x(t)$ is orbitally stable if there is a number $\epsilon > 0$ such that whenever $|y_0 - x(0)| < \epsilon$ then $|y(t,y_0) - x(t + \delta)| \to 0$ as $t \to \infty$ for some number $\delta (0 \leq \delta < T)$.

The eigenvalue problem for the Floquet exponents corresponding to (5.4.4) is

$$\frac{\partial w}{\partial t} + L(\lambda)w + \epsilon M(v,w) - \beta w = 0 \tag{5.4.5}$$

$$w(0) = w(\tfrac{2\pi}{\omega})$$

Recall that $\lambda = \lambda_0 + \tau(\epsilon)$, $\omega(\epsilon)$ and $v(\epsilon,s)$ are predetermined. We change variables by putting $s = \omega t$ and replace β by $\omega\beta$; then we get

$$\omega \frac{\partial w}{\partial s} + (L_0 + \tau L_1)w + \epsilon M(v,w) - \beta\omega w = 0$$

$$w(0) = w(2\pi) \quad . \tag{5.4.6}$$

One solution of (5.4.6) is $\beta = 0$ and $w = \frac{\partial v}{\partial s}$; this solution is obtained by differentiating (5.1.2) with respect to s . When $\epsilon = 0$, $\omega = 1$ and $\lambda = \lambda_0$ we have the eigenvalue problem

$$\frac{\partial w}{\partial s} + L_0 w - \beta w = 0$$

$$w(0) = w(2\pi)$$

Solutions of this problem are of the form $w(x,s) = e^{\beta s}\varphi(x)$, where $\beta = \mu + 2\pi in$ φ is an eigenfunction of L_0 with eigenvalue μ , and n is an arbitrary integer. Under the bifurcation assumption that L_0 has eigenvalues at $\pm i$ and all other eigenvalues in the right half plane, we see that when $\epsilon = 0$ (5.4.6) has double eigenvalues at $\pm 2\pi in$ and all remaining eigenvalues in the right half plane.

As we have seen, one Floquet exponent of (5.4.6) is always $\beta = 0$ with $w = \frac{\partial u}{\partial s}$ (for small $\epsilon \neq 0$) . We construct the other Floquet function in the form

$$\varphi = a(\epsilon) \frac{\partial v}{\partial s} + \psi \quad . \tag{5.4.7}$$

Recall that $u = \epsilon v$ and $v = v_0 + \epsilon v_1 + \ldots = u_0 + \varphi$. The function $\frac{\partial v}{\partial s}$ satisfies

$$\omega \frac{\partial^2 v}{\partial s^2} + L(\lambda) \frac{\partial v}{\partial s} + \epsilon M(v, \frac{\partial v}{\partial s}) = 0 . \qquad (5.4.8)$$

Substituting the expression (5.4.7) for φ into (5.4.6) and using (5.4.8), we get

$$\omega \frac{\partial \psi}{\partial s} + L(\lambda)\psi + \epsilon M(v,\psi) - \beta\omega\psi - \beta\omega a \frac{\partial v}{\partial s} = 0 ,$$

$$\psi(0) = \psi(2\pi) \qquad (5.4.9)$$

We seek to determine $\psi(s,\epsilon)$, $\beta(\epsilon)$ and $a(\epsilon)$ in the form

$$\psi = u_0 + \epsilon \eta , \quad [\eta] = 0$$

$$\beta(\epsilon) = \epsilon^2 \sigma(\epsilon) , \quad \sigma = \sigma_0 + \epsilon \sigma_1 + \ldots$$

$$a(\epsilon) = a_0 + \epsilon a_1 + \ldots \quad .$$

Substituting these series into (5.4.9) we obtain

$$\{J\eta + M(u_0,u_0)\} + \epsilon\{\omega_2 \frac{\partial u_0}{\partial s} + \tau_2 L_1 u_0$$

$$+ M(u_0,\eta) + M(\phi,u_0) - \sigma_0 u_0 - a_0\sigma_0 \frac{\partial u_0}{\partial s} \} + O(\epsilon^2) = 0 \quad .$$

Applying the reduction arguments as in § 5.3 , we obtain

$$\eta + KM(u_0,u_0) + \epsilon\, KQ\{\omega_2 \frac{\partial u_0}{\partial s} + \ldots\} + O(\epsilon^2) = 0$$

$$\tau_2\, Re[L_1u_0] + Re[M(u_0,\eta)] + Re[M(v_1,u_0)] - \sigma_0 + O(\epsilon) = 0$$

$$-\omega_2 + \tau\, Im[L_1u_0] + Im[M(u_0,\eta)] + Im[M(\phi,u_0)] + a_0\sigma_0 + O(\epsilon) = 0 \ .$$

Note that $M(u_0,u_0)$ lies in the range of J by Ex. 1, § 5.4 . When $\epsilon = 0$ the following equations must hold:

$$J\eta_0 + M(u_0,u_0) = 0$$

$$-\sigma_0 + \tau_2\, Re[L_1u_0] + Re[M(u_0,\eta_0)] + Re[M(v_1,u_0)] = 0$$

$$a_0\sigma_0 + \tau_2\, Im[L_1u_0] + Im[M(u_0,\eta_0)] + Im[M(v_1,u_0)] - \omega_2 = 0 \ .$$

Assuming these equations to be satisfied, we now apply an implicit function argument as in § 5.3 . The Frechet derivative of the above equations at $\epsilon = 0$ is

$$\begin{bmatrix} I & 0 & 0 \\ Re[M(u_0,\cdot)] & -1 & 0 \\ Im[M(u_0,\cdot)] & a_0 & \sigma_0 \end{bmatrix} \qquad (5.4.10)$$

(The notation $Re[M(u_0,\cdot)]$ denotes the linear functional $\eta \rightarrow Re[M(u_0,\eta)]$ defined on the Banach space $B = \{\eta : \eta \in P_{2\pi}^{2+2\alpha,\ 1+\alpha}\ ,\ [\eta] = 0\}$.) The Frechet derivative (5.4.10) is invertible provided $\sigma_0 \neq 0$. By a calculation left as an exercise, one obtains

$$\sigma_0 = 2\tau_2\, Re[L_1u_0] \ ; \qquad (5.4.11)$$

hence $\sigma_0 \neq 0$ provided $\tau_2 \neq 0$. In this case, by virtue of the implicit function theorem, we see that there exists an analytic one-parameter family $w(s,\epsilon)$, $a(\epsilon)$, and $\beta(\epsilon)$ satisfying the eigenvalue problem (5.4.6). We have thus constructed the second critical Floquet eigenfunction and eigenvalue at $\beta = 0$. The others, at $\beta = \pm 2\pi in$, may be treated in a similar manner. From (5.4.11) one may conclude, formally, that supercritical periodic solutions are stable and subcritical solutions are unstable. (provided, of course, that $\tau_2 \neq 0$) Recent work by Iooss and Judovic (discussed in the next section) goes toward establishing the validity of the Floquet analysis in determining the stability of the periodic solutions.

<u>Exercises</u>

1. Show that $[N(u_0,u_0)] = 0$.

2. Let $v = u_0 + \epsilon v_1 + \epsilon^2 v_2 + \dots$, $\omega = 1 + \omega_1 \epsilon + \omega_2 \epsilon^2 + \dots$, $\tau = \epsilon \tau_1 + \epsilon^2 \tau_2 + \dots$. Derive the recursion equations determining the coefficients. In particular, show that $\omega_1 = \tau_1 = 0$,

$$Jv_1 + N(u_0,u_0) = 0 ;$$

and calculate τ_2 and ω_2. Show that $\tau_2 \operatorname{Re}[L_1 u_0] + \operatorname{Re}[M(u_1,u_0)] = 0$.

3. Let v be the periodic solution and let η be the function constructed in § 5.4 ($\varphi = a(\epsilon)\frac{\partial v}{\partial s} + \psi$, $\psi = u_0 + \epsilon \eta$). Show that $\eta_0 = 2v_1$ (see exercise 2) Hint: η_0 satisfies $J\eta_0 + M(u_0,u_0) = 0$, $[\eta] = 0$.

4. Derive (5.4.11) and conclude that supercritical periodic solutions are stable and subcritical periodic solutions are unstable.

5.5 Appendix.

Here we indicate the derivation of the estimate (5.2.5). We first note the integral representation

$$e^{-tA}u = \frac{1}{2\pi i} \int_C e^{-\lambda t}(\lambda - A)^{-1}u \, d\lambda \tag{5.5.1}$$

where the contour C extends to infinity in the right half plane along the rays $|\lambda| e^{\pm i\theta}$ $(0 < \theta < \pi/2)$. The desired inequality follows from the estimate $|(\lambda - A)^{-1}f|_{2+2\alpha} \leq c(|\lambda|^2 + 1)|f|_{2\alpha}$ for the resolvent transformation $(\lambda - A)^{-1}$ whose derivative follows below.

Letting $(\lambda - A)u = f$, we have, on taking the inner product with u,

$$\lambda |u|^2 + \|u\|^2 = (f,u)$$

where $\|u\|^2 = (Au,u)$. Taking imaginary parts of this identity and applying Schwartz's inequality, we obtain $|u| \leq |\text{Im } \lambda|^{-1} |f|$. Thus

$$|(\lambda - A)^{-1}f| \leq \frac{1}{|\lambda| \, |\sin \theta|} |f|$$

for $\lambda = |\lambda| e^{\pm i\theta}$. The norm $|u|$ denotes, of course, the L_2 norm. Now we write

$$Au = \lambda u - f \tag{5.5.2}$$

and apply the *a priori* estimate of Solonnikov (Ladyzhenskaya, Theorem 3, p. 78) to get $|u|_{W_{2,2}} \leq c(|\lambda| |u| + |f|) \leq c'|f|$, where $| \ |_{W_{2,2}}$ is the Sobolev norm.

From the Sobolev inequality $|u|_6 \leq c|u|_{W_{1,2}}$ we get that $\nabla u \in L_6$. From the embedding lemma (Theorem 2.2.1), $u \in C_\alpha$, where $\alpha = 1/2$. It follows that $|u|_{1/2} \leq c|u|_{W_{2,2}} \leq c'|f|$. Going back to (5.5.2) we apply the Solonnikov estimates for the Hölder norms to obtain $|u|_{2+1/2} \leq c(|\lambda||u|_{1/2} + |f|_{1/2}) \leq c'(|\lambda| + 1)|f|_{1/2}$. If the Hölder exponent 2α is greater than $1/2$ we repeat the ladder argument one more time.

Notes

The bifurcation of periodic motions from a stationary equilibrium was investigated completely by E. Hopf [1] for ordinary differential equations. Hopf established the exchange of stabilities for this problem. (That is, he showed that supercritical motions are stable while subcritical motions are unstable.)

The bifurcation of periodic solutions of the Navier-Stokes equations has been treated independently, and by different methods, by Sattinger [2], Iooss [2], Judovic [5], and Joseph and Sattinger. A closely related problem, the bifurcation of traveling wave disturbances in plane Poiseuille flow, was treated formally by J. T. Stuart and J. Watson. Numerical work by Reynolds and Potter shows the bifurcation is subcritical. The work of these authors establishes the bifurcation of traveling wave disturbances along an infinite axis. If one fixes one's attention to a single period cell, one would observe time periodic oscillations.

Numerical work by Lee and Luss [2] indicates the appearance of time periodic solutions of the equations (1.2.3) governing chemical reactions. Lee and Luss consider the equations

$$\frac{\partial z}{\partial t} = \Delta z - r(z,y)$$

$$L\frac{\partial y}{\partial t} = \Delta y + \beta r(z,y)$$

$$z = y = 1 \quad \text{on} \quad \partial D$$

The number L is called the Lewis number. For $L = 1$ there is a unique stable stationary state. Their work indicates that as L decreases the stationary state loses equilibrium by a pair of complex conjugate eigenvalues crossing the imaginary axis. They obtain (numerically) a pattern of periodic oscillations.

Ruelle and Takens have treated the problem of the bifurcation of invariant curves of a diffeomorphism which leaves the origin fixed. They consider diffeomorphisms $\Phi(x,\mu)$ with the property that $\Phi(0,\mu) \equiv 0$, $\Phi_x'(0,\mu)$ has a pair of complex conjugate eigenvalues which cross the unit circle as μ crosses 0. Under these, and certain technical assumptions, a closed curve of points, invariant under the action of Φ, bifurcates from the origin. This result is important in trying to deal with the higher bifurcations which take place in a dynamical system, for example, the bifurcation of an invariant torus from a periodic solution when the latter loses stability.

VI

The Mathematical Problems of Hydrodynamic Stability

6.1 Introduction.

In the two preceeding chapters we have assumed the validity of Lyapounov's theorem for the case of the Navier-Stokes equations — that is, that the stability of a stationary solution is determined by the spectrum of the derived operator. What does this assumption involve in the case of the Navier-Stokes equations? Let $\tilde{u}_i$, $\tilde{p}$ be the stationary solutions of the non-linear boundary value problem

$$\frac{1}{R}\,\Delta\tilde{u}_i - \frac{\partial p}{\partial x_i} = \tilde{u}_j\,\frac{\partial\tilde{u}_i}{\partial x_j}$$

$$\frac{\partial\tilde{u}_i}{\partial x_i} = 0 \qquad \tilde{u}_i\Big|_{\partial D} = \psi_i \quad .$$

The (time dependent) perturbed velocity field and pressure, denoted by u_i, p then satisfy the system of equations

$$\frac{\partial u_i}{\partial t} - \frac{1}{R}\,\Delta u_i + \tilde{u}_j\,\frac{\partial u_i}{\partial x_j} + u_j\,\frac{\partial\tilde{u}_i}{\partial x_j} + u_j\,\frac{\partial u_i}{\partial x_j} = -\frac{\partial p}{\partial x_i} \tag{6.1.1}$$

$$\frac{\partial u_i}{\partial x_i} = 0\ , \qquad u_i\Big|_{\partial D} = 0 \quad .$$

By applying the projection P_σ , as before, we can write these equations in the form

$$\frac{\partial u}{\partial t} + Lu + N(u,u) = 0 \tag{6.1.2}$$

where u is a vector field in H_σ and

$$Lu = P_\sigma[-\frac{1}{R}\Delta u + \tilde{u}\cdot\nabla u + u\cdot\nabla\tilde{u}]$$

$$N(u,u) = P_\sigma(u\cdot\nabla u) \quad .$$

We thus reduce the question of stability of $\tilde{u}$ to the asymptotic behavior of the solutions of (6.1.2).

Let us introduce the norms

$$|u|^2 = \int_D u_i u_i \, d\underline{x}$$

$$\|u\|^2 = \int_D \frac{\partial u_i}{\partial x_j}\frac{\partial u_i}{\partial x_j}\, d\underline{x} \quad .$$

Definition: We say that $\tilde{u}$ is stable if given any $\epsilon > 0$ there is a $\delta > 0$ such that whenever $u(t)$ satisfies (6.1.2) and $|u(0)| < \delta$, then $|u(t)| < \epsilon$ for all $t > 0$. We say that $\tilde{u}$ is asymptotically stable if $|u(t)| \to 0$ as $t \to \infty$ whenever $|u(0)| < \delta$; and we say that $\tilde{u}$ is unconditionally stable if it is stable to disturbances of arbitrarily large initial size.

In the case of ordinary differential equations one analyzes the linearized equations

$$\frac{\partial u}{\partial t} + Lu = 0 \qquad (6.1.3)$$

rather than (6.1.2), arguing that, if the perturbations are small, then the nonlinear terms are of second order magnitude and can be neglected. Lyapounov's theorem (Theorem 1.1.1) justifies this argument for the case of ordinary differential equations. In the present case, however, the nonlinear term $N(u,u)$ involves derivatives of u which may not be small. Nevertheless, the "linearization hypothesis" (Lyapounov's first criterion) does hold for (6.1.2) and we shall demonstrate its validity in what follows.

6.2. Energy Methods.

First it will be instructive to give a brief account of energy methods in hydrodynamic stability problems. If we take the inner product of (6.1.2) with u we obtain the differential equation

$$\frac{\partial}{\partial t} \frac{|u|^2}{2} + (Lu,u) = 0 \tag{6.2.1}$$

since $(N(u,u),u) = 0$ (see exercise 1). Now let

$$\gamma = \inf \frac{(Lu,u)}{|u|^2} \tag{6.2.2}$$

subject to the constraints $\operatorname{div} u = 0$ and $u|_{\partial D} = 0$. This variational problem is similar to the classical variational characterization of the first eigenvalue of the Laplacian. From (6.2.1) and (6.2.2) we obtain

$$\frac{\partial}{\partial t} \frac{|u|^2}{2} + \gamma |u|^2 \leq 0 \tag{6.2.3}$$

Integrating this differential inequality we get $|u(t)| \leq |u(0)| e^{-\gamma t}$. Thus, if $\gamma > 0$ all perturbations decay exponentially to zero in the L_2 norm, regardless of the initial size of the disturbance, and the flow $\tilde{u}$ is unconditionally stable.

6.3 Conditional Stability.

Since the operator L is not symmetric it is possible for γ to be negative while all the eigenvalues of L have positive real parts. (see exercise 2) This latter condition, as we shall see, is sufficient to guarantee stability with respect to sufficiently small disturbances. We are going to prove

Theorem 6.3.1. *Let all eigenvalues of L, that is, of the eigenvalue problem*

$$- \frac{1}{R} \Delta\varphi_i + \tilde{u}_j \frac{\partial\varphi_i}{\partial x_j} + \varphi_j \frac{\partial\tilde{u}_i}{\partial x_j} = \lambda\varphi_i - \frac{\partial p}{\partial x_i}$$

$$\frac{\partial\varphi_i}{\partial x_i} = 0 \qquad \varphi_i \Big|_{\partial D} = 0 \tag{6.3.1}$$

have positive real parts. Then $\tilde{u}$ is conditionally asymptotically stable in the L_2 norm.

The actual value of R is immaterial to the following discussion and so may be taken equal to unity from now on. We first need the following two lemmas.

Lemma 6.3.2. *The spectra of L and L^* each consist of a countably infinite set of eigenvalues tending to infinity in the right half plane. Each eigenvalue is of finite multiplicity, and the eigenvalues cannot cluster at any finite complex number. The eigenfunctions of L and also those of L^* are complete in H_σ. (That is, the class of finite linear combinations is dense in H_σ.)*

A proof of lemma 6.3.2 can be found in Sattinger [1]. The Riesz-Schauder theory applies to L and L^*. The eigenvalues of L and of L^*

appear in complex conjugate pairs; and $\bar{\lambda}$ is an eigenvalue of L^* if and only if λ is an eigenvalue of L. For each integer n we let $\mathfrak{M}_n$ be the invariant subspace of L corresponding to those eigenvalues λ with $\mathrm{Re}\,\lambda \leq n$. $\mathfrak{M}_n$ may be characterized as the range of the projection

$$P_n = \frac{1}{2\pi i} \int_{C_n} (\lambda - L)^{-1}\, d\lambda$$

where the contour C_n encloses precisely those eigenvalues λ with $\mathrm{Re}\,\lambda \leq n$. (Dunford & Schwartz pp. 566-77). The adjoint operator P_n^* is the projection

$$P_n^* = \frac{1}{2\pi i} \int_{C_n} (\lambda - L^*)^{-1}\, d\lambda$$

onto the finite dimensional invariant subspace $\mathfrak{M}_n^*$ of L^*. Let $\mathfrak{N}_n$ denote the complementary subspace; $\mathfrak{N}_n$ is the null space of P_n. We have further $\mathfrak{N}_n^\perp = \mathfrak{M}_n^*$. (see exercise 4).

<u>Lemma 6.3.3</u>: <u>Let $\mathfrak{M}_n$ be the invariant subspace of L corresponding to the eigenvalues $\mathrm{Re}\,\lambda \leq n$ and let $\mathfrak{N}_n$ be the complementary subspace. Then for sufficiently large n, $(Lu,u) \geq \frac{1}{2}\|u\|^2$ for $u \in \mathfrak{N}_n$.</u>

<u>Proof</u>: We have

$$\mathrm{Re}(Lu,u) \geq \|u\|^2 - c|u|^2 = \|u\|^2\left(1 - c\,\frac{|u|^2}{\|u\|^2}\right) ,$$

so it suffices to show that $\|u\|^2 \leq 2c|u|^2$ for $u \in \mathfrak{N}_n$ and n sufficiently large. For any subspace $\mathfrak{M}$, let

$$d(\mathfrak{M}) = \inf \frac{\|u\|^2}{|u|^2}$$

subject to $\operatorname{div} u = 0$, $u|_{\partial D} = 0$, and $u \in \mathfrak{M}^{\perp}$. Let $\mu_1, \mu_2, \ldots$ denote the eigenvalues of the operator $Au = P_\sigma(-\Delta u)$. These tend to infinity, and the corresponding eigenfunctions φ_i form an orthonormal sequence. (see Ladyzhenskaya, p. 44). We shall prove: for any k there exists an n sufficiently large such that $d(\mathfrak{M}_n^*) \geq \mu_k$. Since $\mathfrak{M}_n^{*\perp} = \mathfrak{N}_n$ this is sufficient to establish the lemma.

The eigenvalue μ_k has the variational characterization

$$\mu_k = \inf \frac{\|u\|^2}{|u|^2}$$

subject to $\operatorname{div} u = 0$, $u|_{\partial D} = 0$, and $(u, \varphi_i) = 0$ for $i = 1, \ldots, k-1$, where $\varphi_1, \ldots \varphi_{k-1}$ are the first $(k-1)$ eigenfunctions. Let E_k denote the orthogonal projection onto the subspace spanned by $\{\varphi_1, \ldots, \varphi_{k+1}\}$, and let $F_k = I - E_k$. Then, as is easily seen,

$$|E_k u|^2 + |F_k u|^2 = |u|^2 ,$$

$$\|E_k u\|^2 + \|F_k u\|^2 = \|u\|^2 , \tag{6.3.2}$$

and

$$\|F_k u\|^2 \geq \mu_{k+1} \, |F_k u|^2$$

Since the eigenfunctions of L^* are complete, there are, for fixed $\epsilon > 0$, fixed k , and sufficiently large n , functions $v_1, \ldots, v_{k+1}$ in $\mathfrak{M}_n^*$ such that $|\varphi_i - v_i| < \epsilon^{1/2}(k+1)^{-1/2}$. Now if u is orthogonal to $\mathfrak{M}_n^*$,

$$|(u,\varphi_i)| = |(u,\varphi_i - v_i)| \leq \frac{\epsilon^{1/2}|u|}{\sqrt{k+1}} .$$

Consequently,

$$|E_k u|^2 = \sum_{i=1}^{k+1} |(u,\varphi_i)|^2 \leq \epsilon\, |u|^2$$

and

$$|F_k u|^2 \geq (1 - \epsilon)\, |u|^2 .$$

On the other hand, if $\|u\| < +\infty$ we have

$$\|u\|^2 = \|E_k u\|^2 + \|F_k u\|^2 \geq \|F_k u\|^2 \geq \mu_{k+1}\, |F_k u|^2 \geq \mu_{k+1}(1 - \epsilon)|u|^2$$

Therefore, for n sufficiently large, $d(\mathfrak{M}_n^*) \geq \mu_k$, and the proof of the lemma is complete.

Lemma 6.3.4. Let $\mathfrak{M}$ be a finite dimensional subspace of H_σ spanned by $\psi_1, \ldots, \psi_n$. Let $u \in \mathfrak{M}$ have the expansion

$$u = \sum_k c_k \psi_k .$$

Then: (i) there are constants d and d' such that

$$d\Big(\sum_k c_k^2\Big)^{1/2} \leq |u| \leq d'\Big(\sum_k c_k^2\Big)^{1/2} .$$

(ii) If $\|\psi_i\| < +\infty$ and if $\psi_i|_{\partial D} = 0$ for $i=1, \ldots, n$, then the norms $|\ |$ and $\|\ \|$ are equivalent on $\mathfrak{M}$; that is, there are constants d and d' such that $d|u| \leq \|u\| \leq d'|u|$ for all u in $\mathfrak{M}$.

<u>Proof</u>: Statement (i) is a simple matter of linear algebra. We have

$$|u| \leq \sum_k |c_k||\psi_k| \leq (\sum_k |c_k|^2)^{1/2} (\sum_k |\psi_k|^2)^{1/2}$$

so we may take

$$d' = (\sum_k |\psi_k|^2)^{1/2} .$$

On the other hand, the c_k are given as solutions of the system of equations

$$(u, \psi_j) = \sum_k (\psi_k, \psi_j) c_k = A_{jk} c_k$$

The matrix A_{jk} is invertible since the $\{\psi_j\}$ are linearly independent. Therefore $c_k = (A_{jk})^{-1}(u, \psi_j)$ and the remaining estimate follows immediately.

The first inequality in (ii), namely $\|u\| \geq d|u|$ is simply a consequence of the variational inequality $\|u\|^2 \geq \gamma |u|^2$ which holds on the entire space. To get the second half we employ the Gram Schmidt process to replace the $\{\psi_i\}$ by an orthonormal sequence $\{\varphi_i\}$ (in the inner product for H_σ). Given u in $\mathfrak{M}$ we write

$$u = \sum_k (u, \varphi_k)\varphi_k = \sum_k a_k \varphi_k .$$

Then $|u|^2 = \sum_k |(u, \varphi_k)|^2 = \sum_k a_k^2$, while

$$\|u\|^2 = \int_D \sum_{k,\ell} a_k a_\ell \frac{\partial(\varphi_k)_i}{\partial x_j} \frac{\partial(\varphi_\ell)_i}{\partial x_j} d\underline{x}$$

$$\leq \sum_{k,\ell} a_k a_\ell \|\varphi_k\| \|\varphi_\ell\| \leq c \sum_k a_k^2 = c|u|^2 .$$

Here $(\varphi_k)_i$ denotes the ith component of φ_k .

6.4. <u>The stability theorem</u>:

We now prove Theorem 6.3.1. Let all eigenvalues of L lie to the right of $\operatorname{Re} \lambda = \gamma > 0$. Choose subspaces $\mathfrak{M}_n$ and $\mathfrak{N}_n$ as in Lemma 6.3.3. We write $u = u_1 + u_2$ where $u_1 \in \mathfrak{M}_n$ and $u_2 \in \mathfrak{N}_n$. From (6.2.1) we have

$$\frac{d}{dt} \frac{|u|^2}{2} + (Lu_2,u_2) = - (Lu_1,u_2) - (Lu_2,u_1) - (Lu_1,u_1) \quad (6.4.1)$$

The terms on the right may be estimated as follows. When L is restricted to the finite dimensional subspace $\mathfrak{M}_n$ it becomes a bounded operator; thus there is a constant $c_1 = c_1(n)$ such that

$$|Lu_1| \leq c_1|u_1|$$

and

$$|(Lu_1,u_2)| \leq |Lu_1| \cdot |u_2| \leq c_1|u_1| \cdot |u_2| \quad ,$$

$$|(Lu_1,u_1)| \leq |Lu_1| \cdot |u_1| \leq c_1|u_1|^2 \quad .$$

We treat the remaining term as follows:

$$(Lu_2,u_1) = (u_2,L^* u_1) = (u_2,Lu_1) + (u_2,(L^* - L)u_1)$$

Since $L^* - L$ is a first order partial differential operator, $|(L^* - L)u_1| \leq c\|u_1\| \leq c_2|u_1|$ by Lemma 6.3.4 (ii). It follows that

$$|(Lu_2,u_1)| \leq |u_2|\cdot|Lu_1| + c_2|u_2|\cdot|u_1| \leq (c_1 + c_2)|u_1|\cdot|u_2| \quad .$$

Using these estimates, Lemma 6.3.3, (6.4.1), and the inequality

$$ab \leq \frac{a^2}{4\epsilon} + \epsilon b^2 \quad ,$$

we obtain

$$\frac{d}{dt}\frac{|u|^2}{2} + \frac{\|u_2\|^2}{2} \leq c_3|u_1|^2 + \epsilon\,|u_2|^2 \quad . \tag{6.4.2}$$

Here the constant $c_3 = c_3(\epsilon, n)$ tends to infinity as $n \to \infty$ or as $\epsilon \to 0$.

By choosing n sufficiently large we can guarantee that $\|u_2\|^2 \geq c_4|u_2|^2$, where c_4 is as large as we please. Of course, choosing n large results in a large value for c_3 , but some constants are more important than others. So, having done this, we get

$$\frac{d}{dt}\frac{|u|^2}{2} + c_4|u_2|^2 \leq c_3|u_1|^2 \tag{6.4.3}$$

(We can neglect the term $\epsilon\,|u_2|^2$ on the right side of 6.4.2 by incorporating it in the term $c_4|u_2|^2$ on the left.) Since $u = u_1 + u_2$, $|u|^2 \leq 2|u_1|^2 + 2|u_2|^2$, so

$$|u_2|^2 \geq \frac{|u|^2}{2} - |u_1|^2 \quad ;$$

inserting this in (6.4.3) we obtain

$$\frac{d}{dt}|u|^2 + c_4|u|^2 \leq c_5|u_1|^2 \tag{6.4.4}$$

where, again, c_4 and c_5 grow with n .

The next step is to estimate $|u_1|$. We return to the equations for the perturbations, <u>viz</u>.

$$\frac{du}{dt} + Lu + N(u,u) = 0 \quad ,$$

and operate with the projection P_n . The projection P_n is of the form

$$P_n u = \sum_j (u, \varphi_j^*) \varphi_j$$

where the $\{\varphi_j\}$ form a Jordan basis for L on $\mathfrak{M}_n$ and the $\{\varphi_j^*\}$ form a Jordan basis for L^* on $\mathfrak{M}_n^*$. Since $u_1 = P_n u$ we get

$$\frac{du_1}{dt} + Lu_1 + P_n N(u,u) = 0 \quad .$$

Let us put $q_j(t) = (u, \varphi_j^*)$; then this equation is equivalent to the system of ordinary differential equations

$$\frac{dq_k}{dt} + L_{kj} q_j + (N(u,u), \varphi_k^*) = 0 \qquad (6.4.5)$$

where L_{ij} is the matrix of L with respect to $\{\varphi_j\}$, viz. $L_{ij} = (L\varphi_i, \varphi_j^*)$. Equations (6.4.5) can be obtained by taking the inner product of the preceeding equation with φ_i^* . (The φ_i, φ_j^* satisfy $(\varphi_i, \varphi_j^*) = \delta_{ij}$) .

Now $N(u,u) = P_\sigma(u \cdot \nabla u)$ and φ_k^* is a divergence-free vector field, so

$$\begin{aligned}(N(u,u), \varphi_k^*) &= (P_\sigma\, u \cdot \nabla u, \varphi_k^*) \\ &= (u \cdot \nabla u, \varphi_k^*) \\ &= \int_D u_i \frac{\partial u_i}{\partial x_j} \overline{(\varphi_k^*)_i} \\ &= \int_D \frac{\partial}{\partial x_j} u_i u_j \overline{(\varphi_k^*)_i} \\ &= -\int_D u_i u_j \frac{\partial}{\partial x_j} \overline{(\varphi_k^*)_i} \quad .\end{aligned}$$

Consequently, assuming the φ_k^* have bounded first derivatives,

$$|(N(u,u), \varphi_k^*)| \leq c_6 \; |u|^2$$

where, as usual, $c_6 = c_6(n) \to \infty$ as $n \to \infty$.

Since the eigenvalues of the matrix L_{ij} are simply those of L restricted to the invariant subspace $\mathfrak{M}_n$, these all have positive real parts $\operatorname{Re} \lambda > \gamma$. We therefore obtain, by methods of ordinary differential equations, the following estimate:

$$q(t) \leq c_6 e^{-\gamma t} \, q(0) + c_7 \int_0^t e^{-\gamma(t - s)} |u(s)|^2 ds \; , \tag{6.4.6}$$

where

$$q(t) = (\sum_i q_i^2)^{1/2} \quad .$$

From lemma (6.3.4) we see that the norm q is equivalent to $|u_1|$; therefore, in (6.4.6), q may be replaced everywhere by $|u_1|$. Squaring this inequality we obtain

$$|u_1(t)|^2 \leq 2c_6^2 \; e^{-2\gamma t} \; |u_1(0)|^2$$

$$+ \; 2c_7^2 \left[\int_0^t e^{-\gamma(t - s)} \; ds \right] \left[\int_0^t e^{-\gamma(t - s)} \; |u(s)|^4 ds \right] \quad .$$

Since $|u_1(0)| = |P_n u(0)| \leq c|u(0)|$, and since

$$\left| \int_0^t e^{-\gamma(t - s)} \; ds \right| \leq \text{ const}$$

as $t \to \infty$, there are constants c_8 and c_9 such that

$$|u_1(t)|^2 \le c_8 e^{-2\gamma t} |u(0)|^2 + c_9 \int_0^t e^{-\gamma(t-s)} |u(s)|^4 \, ds .$$

We now insert this estimate into the right side of (6.4.4), obtaining

$$\frac{d}{dt} |u|^2 + c_4 |u|^2 \le c_{10} e^{-2\gamma t} |u(0)|^2 + c_{11} \int_0^t e^{-\gamma(t-s)} |u(s)|^4 \, ds .$$

We integrate this differential inequality, obtaining

$$|u(t)|^2 \le c_{12} e^{-c_4 t} |u(0)|^2 + c_{11} e^{-c_4 t} \int_0^t \int_0^\tau e^{c_4 \tau - \gamma(\tau - s)} |u(s)|^4 \, ds d\tau .$$

Recall that c_4 can be made large, and in particular we can assume that $c_4 > 2\gamma$. We now interchange the order of integration in the iterated integral. We get

$$e^{-c_4 t} \int_0^t \int_0^\tau e^{(c_4 - \gamma)\tau} e^{\gamma s} |u(s)|^4 ds d\tau$$

$$= e^{-c_4 t} \int_0^t e^{\gamma s} |u(s)|^4 ds \int_s^t e^{(c_4 - \gamma)\tau} d\tau$$

$$= \int_0^t \frac{e^{-\gamma(t-s)} - e^{-c_4(t-s)}}{c_4 - \gamma} |u(s)|^4 \, ds$$

$$\le c \int_0^t e^{-\gamma(t-s)} |u(s)|^4 \, ds$$

since $(c_4 - \gamma) > 0$. We thus finally obtain the integral inequality

$$|u(t)|^2 \le c_{12} e^{-\gamma t} |u(0)|^2 + c_{13} \int_0^t e^{-\gamma(t-s)} |u(s)|^4 \, ds$$

for $|u(t)|^2$. Now setting $\alpha(t) = e^{\gamma t} |u(t)|^2$ the above integral inequality becomes

$$\alpha(t) \le c_{12} \, \alpha(0) + c_{13} \int_0^t e^{-\gamma s} \alpha^2(s) ds$$

However, the solution of this integral inequality is dominated by the solution $\beta(t)$ of

$$\beta(t) = c_{12}\alpha(0) + c_{13}\int_0^t e^{-\gamma s}\,\beta^2(s)ds \tag{6.4.7}$$

Differentiating (6.4.7) we obtain the equation

$$\beta' = c_{13}e^{-\gamma s}\,\beta^2(s),\ \beta(0) = c_{12}\alpha(0)$$

whose solution is

$$\beta(t) = \frac{c_{12}\,\gamma\,\alpha(0)}{\gamma - c_{13}\,\alpha(0)(1 - e^{-\gamma t})} .$$

Thus

$$|u(t)|^2 \le \frac{c_{12}\,\gamma|u(0)|^2 e^{-\gamma t}}{\gamma - c_{13}\,|u(0)|^2(1 - e^{-\gamma t})} ,$$

and, if $|u(0)|^2 < \gamma/c_{13}$ then $|u(t)| \to 0$ exponentially as $t \to \infty$.

This completes the stability proof.

Exercises.

1. Show $(N(u,u),u) = 0$. Hint: $N(u,u) = P_\sigma(u\cdot\nabla u)$ so you must prove $(P_\sigma u\cdot\nabla u,\ u) = (u\cdot\nabla u,\ P_\sigma u) = (u\cdot\nabla u,\ u) = 0$. This is

$$(u\cdot\nabla u,\ u) = \int_D u_j \frac{\partial u_i}{\partial x_j}\,u_i\,dx .$$

2. Let L be the matrix

$$L = \begin{pmatrix} \epsilon & -1 \\ 0 & \epsilon \end{pmatrix}$$

Show that for ϵ sufficiently small there are vectors x such that $(Lx,\ x) < 0$; yet all eigenvalues of L are positive. (Use the standard inner product.)

3. Use the relationship $R(\lambda,T)^* = R(\bar{\lambda},T^*)$ for an operator T to prove the following:

   (a) λ is an eigenvalue of L if and only if $\bar{\lambda}$ is an eigenvalue of L^*.

   (b) $P_n^* = \frac{1}{2\pi i} \int_{C_n} (\lambda - L^*)^{-1} \, d\lambda$

4. Prove that P_n^* is the projector onto the finite dimensional invariant subspace $\mathfrak{m}_n^*$ and that $\mathfrak{n}_n^{\perp} = \mathfrak{m}_n^*$.
   Hint: $u \in \mathfrak{m}_n^*$ iff $Q^* u = 0$ while $u \in \mathfrak{n}_n^{\perp}$ iff $(u, Q\varphi) = 0$ for all φ.

5. Let

$$x + Lx = 0 \qquad x(0) = x_0$$

where x is an n - vector, and L is an $n \times n$ matrix. If all eigenvalues of L have real parts greater than $\gamma > 0$, show that there exists a constant $K > 0$ such that $|x(t)| \leq Ke^{-\gamma t} |x(0)|$, where $|\ |$ is the Euclidean norm on R^n.
(Hint: put L in Jordan canonical form)

Notes

The energy argument of § 6.2 is discussed by J. Serrin [1]. Energy methods in hydrodynamic stability problems can be carried much further than indicated here. (See D. Joseph [1].) The first general stability theorem in terms of the spectrum of L was proved by G. Prodi in 1962. Prodi proved that if the spectrum of L lies in the right half plane, and if $\|v(0)\|$ is sufficiently small, then $\|v(t)\|$ remains small and tends to zero as $t \to \infty$. Prodi used arguments based on differential and integral inequalities similar to those presented here. A number of stability and instability results were stated by Judovic [1] in 1965, but no proofs were given. Kirchgässner and Sorger [1] first proved an instability theorem along the lines of Prodi's argument in 1968. They proved that if L has a simple eigenvalue in the left half plane, and if $|Av|$ (L_2 norm of Au) remains finite for all time, then the basic flow is unstable.

Stability and instability theorems for solutions of the Navier-Stokes equations in the Hopf class of weak solutions were proved by Sattinger [1].

The stability of time periodic solutions of the Navier-Stokes equations has been investigated by Judovic [3,4] and by Iooss [2].

For further discussion of the mathematical theory of hydrodynamic stability the reader may consult the articles by Iooss [1] and by Kirchgässner and H. Kielhöfer.

VII

Topological Degree Theory and Applications

7.1. Finite dimensional degree theory. The applications of degree theory to problems in bifurcation theory are numerous and varied, and we can only give a few examples here. We begin by summerizing the basic properties of the finite dimensional degree of a mapping. (See Heinz, and J. T. Schwartz)

Let f be a C^1 mapping defined on a bounded region Ω in R^n and let p be a point in R^n with $f \neq p$ on $\partial\Omega$. The degree $d(f,\Omega,p)$ is defined (see § 7.3 for an analytical definition) and has the following properties.

(i) Let Ω_1 and Ω_2 be two disjoint bounded open sets in R^n, with $f(x) \neq p$ for $x \in \partial\Omega_1 \cup \partial\Omega_2$. Then $d(f,\Omega_1 \cup \Omega_2,p) = d(f,\Omega_1,p) + d(f,\Omega_2,p)$.

(ii) If $d(f,\Omega,p) \neq 0$ then there is at least one point $\xi \in \Omega$ such that $f(\xi) = p$.

(iii) Let $f(x,\lambda)$ be continuous on $\Omega \times I$, where $\lambda \in I = [a,b]$, and let $f(x,\lambda) \neq p$ for $x \in \partial\Omega$ and $\lambda \in I$. Then $d(f(x,\lambda), \Omega,p)$ is constant in λ.

(iv) Let f_1 and f_2 be defined on Ω and satisfy

$$|f_1(x) - f_2(x)| < |f_1(x) - p|$$

for $x \in \partial\Omega$. Then $d(f_1,\Omega,p) = d(f_2,\Omega,p)$.

Property (iii) is a statement of the important fact that the degree remains invariant under deformations so long as no solutions cross the boundary. The degree can be extended in a unique way to mappings which are merely continuous by approximating them by differentiable mappings.

The <u>index</u> $i = [f,\xi]$ of f relative to an isolated solution ξ of $f(\xi) = p$ is defined to be

$$i[f,\xi] = d(f,\Omega_0,p)$$

where Ω_0 is a neighborhood of ξ which contains no other solutions of the equation $f(x) = p$.

(v) If $f = p$ has only a finite number of solutions $\xi_1, \ldots, \xi_k$ in Ω , then

$$d(f,\Omega,p) = \sum_{j=1}^{k} i[f,\xi_j] \ .$$

(vi) If f is C^1 and if the Jacobian $\det\left|\frac{\partial f_i}{\partial x_j}\right| \neq 0$ at a solution ξ of $f = p$, then

$$i[f,\xi] = \operatorname{sgn}\det\left.\frac{\partial f_i}{\partial x_j}\right|_{x = \xi} \ .$$

(vii) The degree is always an integer, and the degree of the identity map is 1 .

An elementary analytic derivation of these as well as of other properties of the degree of mapping is given in the article by Heinz and in J. T. Schwartz's notes (see also § 7.3). Let us now use the above properties to give a discussion of bifurcation and stability for the equations

$$f(x,\lambda) = 0 \qquad (7.1.1)$$

where λ is a parameter and $f,x \in R^n$.

Suppose that $x = 0$ is always a solution of (7.1.1). Then 0 is an equilibrium solution of the system

$$\dot{x} = f(x,\lambda) ,$$

whose stability is determined by the eigenvalues of the matrix

$$\left\| \frac{\partial f_i}{\partial x_j} (0,\lambda) \right\| . \tag{7.1.2}$$

Suppose that as λ crosses λ_0 a simple eigenvalue of the matrix (7.1.2) crosses through the origin. If we denote the eigenvalues by $\sigma_1(\lambda), \ldots, \sigma_n(\lambda)$, then, say, for $\lambda < \lambda_0$, Re $\sigma_1(\lambda) < 0$ while $\sigma_1(\lambda)$ crosses through the origin as λ crosses λ_0 . We assume that $\sigma_2, \ldots, \sigma_n$ remain in the left half plane. The null solution thus goes unstable.

Let us calculate the degree

$$d(f,\Omega,0)$$

where $f = f(x,\lambda)$ and Ω is a small neighborhood of the origin. The degree is the sum of the indices of the solutions ξ of $f(\xi,\lambda) = 0$. The index of the null solution is

$$i[f,0] = \text{sgn det} \left| \frac{\partial f_i}{\partial x_j} (0,\lambda) \right|$$

$$= \text{sgn } \sigma_1(\lambda) \ldots \sigma_n(\lambda)$$

Since $\sigma_2, \ldots, \sigma_n$ remain in the left half plane as λ crosses λ_0

while σ_1 changes sign, we see that the index $i[f,0]$ changes sign as λ crosses λ_0. (We are assuming f is real so that all eigenvalues appear in complex conjugate pairs; hence the product $\sigma_1 \ldots \sigma_n$ is real.)

From property (iii) we know that, if no solutions of $f(x,\lambda) = 0$ cross $\partial\Omega$ as λ crosses λ_0, then the equation $f(x,\lambda) = 0$ must have solutions in Ω other than $x = 0$. These may occur either for $\lambda > \lambda_0$ or $\lambda < \lambda_0$ or both. For, if no solutions cross $\partial\Omega$ then $d(f(\cdot,\lambda),\Omega,0)$ remains constant. On the other hand, the index of $x = 0$ changes by a factor of two (from ± 1 to ∓ 1). The only way the degree could remain constant is by the presence of other solutions.

For example, suppose that $x = 0$ is the only solution of $f(x,\lambda) = 0$ for $\lambda < \lambda_0$; and let us say, to be specific, that its index is $+1$ for $\lambda < \lambda_0$ and -1 for $\lambda > \lambda_0$. Then

$$d(f,\Omega,0) = i[f,0] = +1 \quad \text{for } \lambda < \lambda_0$$

and so $d(f,\Omega,0) = +1$ for $\lambda > \lambda_0$ too. But now when $\lambda > \lambda_0$ the index $i[f,0] = -1$; so in order for the degree to be $+1$ there must be other solutions, for example two other solutions each with index $+1$. (Of course, there could also be six other solutions for $\lambda > \lambda_0$ — four with index $+1$ and two with index -1). A bifurcation diagram might thus look like this:

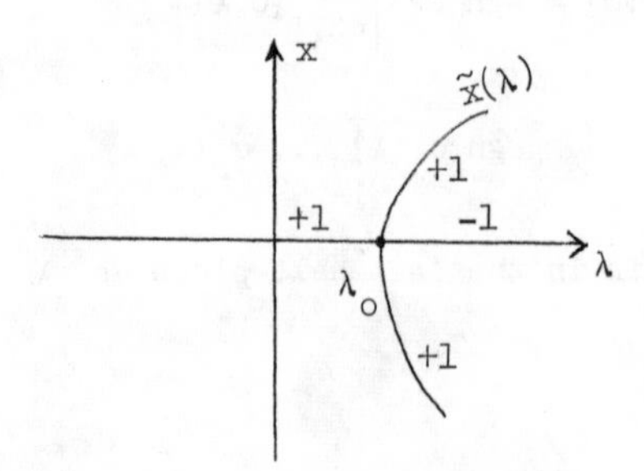

From the degree theory we can sometimes also determine the stability of the bifurcating solutions. Let us suppose that the bifurcation takes place as in the above diagram and denote one of the bifurcating solution branches by $\tilde{x}(\lambda)$. We assume

(i) $\tilde{x}(\lambda) \to 0$ as $\lambda \to \lambda_0^+$

(ii) $\frac{\partial f_i}{\partial x_j}(\tilde{x}(\lambda),\lambda)$ is invertible for $\lambda_0 < \lambda < \lambda_0 + \delta$ for some $\delta > 0$.

Let the eigenvalues of $\tilde{x}(\lambda)$ be denoted by $\tilde{\sigma}_i(\lambda)$, $i = 1, \ldots, n$. Now by degree theory the index $i[f,\tilde{x}]$ must be $+1$ and furthermore, since the eigenvalues of

$$\frac{\partial f_i}{\partial x_j}(\tilde{x},\lambda)$$

vary continuously as $\lambda \to \lambda_0$, we have

$$\tilde{\sigma}_i(\lambda) \to \sigma_i(\lambda_0) \quad \text{as} \quad \lambda \to \lambda_0^+ \ .$$

Therefore, for λ sufficiently close to λ_0 we have

$$\operatorname{sgn} \tilde{\sigma}_2(\lambda) \ldots \tilde{\sigma}_n(\lambda) = \operatorname{sgn} \sigma_2(\lambda) \ldots \sigma_n(\lambda) \ ;$$

and, since

$$\operatorname{sgn} \tilde{\sigma}_1 \ldots \tilde{\sigma}_n = +1$$

while

$$\operatorname{sgn} \sigma_1 \ldots \sigma_n = -1$$

we see that $\tilde{\sigma}_1$ must have the opposite sign to σ_1 . In other words, $\tilde{x}$ is stable.

The above arguments form the basis of the application of topological degree theory to bifurcation problems. In the next sections we shall discuss the extension of degree theory to problems in Banach spaces.

Exercises:

1. Prove the Brouwer fixed point theorem: Let f be a continuous mapping of the closed ball $B = \{x : |x| \leq 1\}$ into itself; then there exists at least one fixed point in B — that is, $f(\xi) = \xi$ for at least one ξ with $|\xi| \leq 1$. Hint: Let $f(x,\tau) = x - \tau f(x)$.

2. Find the indices of the solutions and topological degree of the appropriate maps in the following bifurcation problems

$$\lambda x + ax^n = 0$$

where $a = \pm 1$ and $n = 2,3$. Do the same for the problems

$$x(x^2 - \lambda)^n = 0 \,, \quad n = 1,2,3$$

3. Suppose that $f(0,\lambda) \equiv 0$ and that as λ crosses λ_0 a simple eigenvalue of $\frac{\partial f_i}{\partial x_j}(0,\lambda)$ crosses the origin. Find the indices and determine the stabilities of the bifurcating solutions when the bifurcation diagrams have either of the forms shown below:

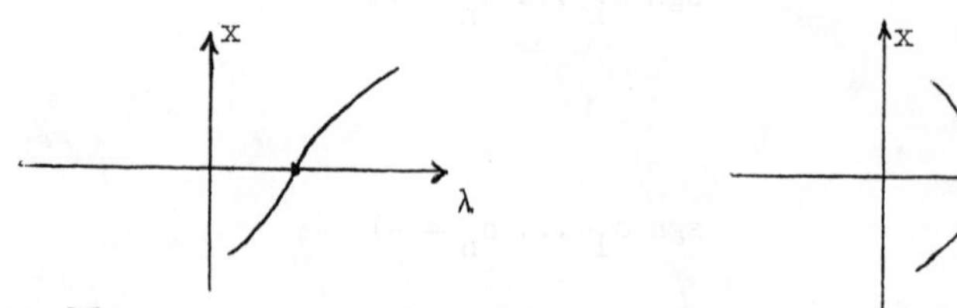

(Assume $\frac{\partial f_i}{\partial x_j}(\tilde{x}(\lambda),\lambda)$ is invertible along the bifurcating curves in a neighborhood of $(\lambda_0,0)$.

4. Construct a mapping $f(x,\lambda)$ whose bifurcation diagram looks like

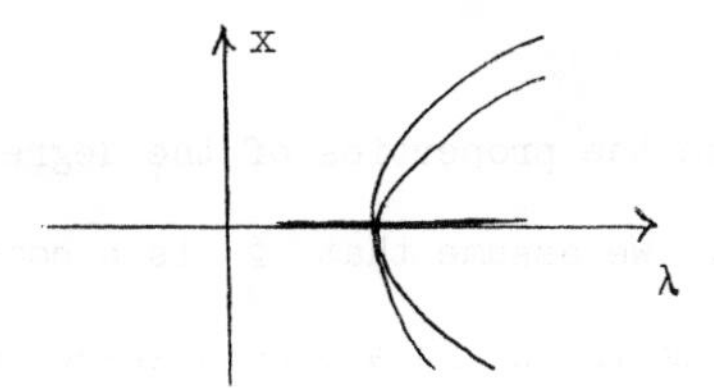

7.2. Leray-Schauder Degree Theory.

In their classic 1934 paper Leray and Schauder extended finite dimensional degree theory to functional equations of the form

$$u - \Phi(\lambda,u) = 0 \qquad (7.2.1)$$

where u lies in a Banach space, Φ is a completely continuous transformation, and λ is a real parameter. Their degree method makes it possible to prove existence theorems for (7.2.1) by simply obtaining a priori estimates for the solutions, thus replacing the so called continuity method.

The continuity method consisted of the following: solve (7.2.1) for a given value of λ, say λ_0, and then continue the solution by successive approximations for other values of λ by applying local existence theorems. This procedure has the disadvantage of requiring that, at each stage of the process, the operator

$$I - \Phi_u'(\lambda,u)$$

be invertible. In particular, the continuity method does not work in case bifurcation takes place somewhere along the solution curve $(\lambda,u(\lambda))$. The degree arguments are not limited by this possibility, and thus provide a more powerful tool for the investigation of solutions of (7.2.1).

In this section we summarize the properties of the degree as developed by Leray and Schauder. We assume that Φ is a completely continuous transformation on a domain Ω in a Banach space B and that Φ is uniformly continuous in λ for u in Ω and $\lambda \in I$. (I is a finite interval [a,b])

(i) If Φ has no fixed points on $\partial\Omega$ then the Leray-Schauder degree $d(I - \Phi,\Omega)$ is defined and is an integer. (A fixed point of Φ is a solution of 7.2.1).

(ii) If, as λ varies from a to b , no fixed points of Φ cross $\partial\Omega$ then $d(I - \Phi(\cdot,a),\Omega) = d(I - \Phi(\cdot,b),\Omega)$.

(iii) If u_0 is an isolated fixed point of Φ in Ω then the index of u_0 , denoted by $i[\Phi,u_0]$, is $d(I - \Phi,\Omega_0)$, where $\Omega_0 \subset \Omega$ is a neighborhood of u_0 containing no other fixed points of Φ .

(iv) If Φ has a finite number of fixed points $u_1, \ldots, u_n$ in Ω then $d(I - \Phi,\Omega) = \sum_{k=1}^{n} i[\Phi,u_k]$.

(v) The degree of the identity map is 1 .

(vi) Let u_0 be a fixed point of Φ and let $A = \Phi'(u_0)$ be the Frechet derivative of Φ at u_0 . Then A is compact. If I - A is invertible then u_0 is an isolated fixed point and $i[\Phi,u_0] = d(I - A,\Omega)$.

(vii) Let μ_0 be a characteristic value of A of multiplicity m . (That is, if $\eta_k = \{\varphi : (I - \mu A)^k\varphi = 0\}$ then $\eta_0 = \{0\} \subset \eta_1 \subset \ldots \subset \eta_m = \eta_{m+1} = \ldots$). Then $d(I - \mu A,\Omega)$ changes by $(-1)^m$ as μ crosses μ_0 .

(viii) If $d(I - \Phi,\Omega) \neq 0$ then Φ has at least one fixed point in Ω .

7.3. <u>Definition of Degree</u>. In this section we give the basic definitions of degree following Heinz for the finite dimensional case and Leray and Schauder for the infinite dimensional case.

The degree of a C^1 mapping in the finite dimensional case is defined as follows. Let $\epsilon = \min_{x \in \partial\Omega} |f(x) - p|$ and let Φ be any positive function on $[0,\infty)$ which vanishes in a neighborhood of $r = 0$ and for $r \geq \epsilon$; Φ is also to satisfy the condition

$$\int_{E^n} \Phi(|x|)dx = 1 .$$

Then

$$d(f,\Omega,p) = \int_{\Omega} \Phi(|f(x) - p|)Jf(x)dx$$

where Jf is the Jacobian $\det \frac{\partial f_i}{\partial x_j}(x)$.

Using this definition it can be shown that the degree is invariant under continuous deformations of f ; thus the degree of a continuous mapping may be defined by approximating it by C^1 mappings.

Let us now define the Leray-Schauder degree for a transformation of the form $I - \Phi$, where Φ is a completely continuous transformation on a Banach space B . For a nonlinear transformation we define complete continuity to mean that $\Phi(\bar{\omega})$ is compact whenever ω is bounded. Suppose ω is a bounded domain and that there are no fixed points of Φ on $\partial\omega$. Then

(1) There is an $\epsilon > 0$ such that $\|u - \Phi(u)\| > \epsilon > 0$ for $u \in \partial\omega$. If not, there would be a sequence $u_n \in \partial\omega$ such that

$y_n = u_n - \Phi(u_n) \to 0$. Since Φ is completely continuous there is a subsequence, which we may take to be u_n itself, such that $\{\Phi(u_n)\}$ is convergent. Then $u_n = y_n + \Phi(u_n)$ is itself convergent, and in the limit we obtain $u = \lim u_n = \Phi(u)$. This contradicts our assumption that Φ has no fixed points on $\partial\omega$.

(2) Since $\Phi(\bar{\omega})$ is compact there is a set of elements $\eta_1, \ldots, \eta_p$ such that for every $x \in \Phi(\bar{\omega})$, $\|x - \eta_i\| < \epsilon/2$ for at least one of the η_i .

(3) Define the transformation T_ϵ on $\Phi(\bar{\omega})$ by

$$T_\epsilon(x) = \frac{\sum_{i=1}^{p} \mu_i(x)\eta_i}{\sum_{i=1}^{p} \mu_i(x)}$$

where

$$\mu_i(x) = \begin{cases} \epsilon/2 - \|x - \eta_i\| & \text{if } \|x - \eta_i\| < \epsilon/2 \\ 0 & \text{if } \|x - \eta_i\| \geq \epsilon/2 \end{cases} .$$

T_ϵ is continuous and its range lies in the finite dimensional subspace spanned by $[\eta_1, \ldots, \eta_p]$. Furthermore, for $x \in \Phi(\bar{\omega})$,

$$\|x - T_\epsilon(x)\| \leq \epsilon/2 .$$

(4) Let $\Phi_\epsilon(x) = T_\epsilon(\Phi(x))$ for $x \in \Omega$. Then $\|\Phi_\epsilon(x) - \Phi(x)\| < \epsilon/2$ for $x \in \omega$ and the range of Φ_ϵ lies in a finite dimensional subspace M . Let N be any finite dimensional subspace containing M and at least one point $x \in \omega$. Let $\omega_N = \omega \cap N$ and consider the mapping

$$F_\epsilon(x) = x - \Phi_\epsilon(x)$$

defined for $x \in \omega_N$. Clearly F_ϵ is a transformation from a subdomain ω_N of N into N . Furthermore, for $x \in \partial\omega_N = \partial\omega \cap N$, we have

$$\|F_\epsilon(x)\| = \|x - \Phi_\epsilon(x)\| = \|(x - \Phi(x)) + (\Phi(x) - \Phi_\epsilon(x)\|$$

$$\geq \|x - \Phi(x)\| - \|\Phi(x) - \Phi_\epsilon(x)\| \geq \epsilon/2 \quad .$$

Therefore $d(F_\epsilon(x),\omega_N,0)$ is defined (relative to N) , and we define

$$d(I - \Phi,\omega) = d(F_\epsilon,\omega_N,0) \quad .$$

Theorem 7.3.1 (Leray,Schauder): The above definition of degree is independent of the choice of the approximation Φ_ϵ and N . Therefore the above procedure yields a unique definition of degree.

Proof: First consider the case of two approximating transformations Φ_ϵ and Ψ_ϵ with the same finite dimensional subspace N . Put

$$U_\theta = \theta\,\Phi_\epsilon + (1 - \theta)\Psi_\epsilon \ , \quad 0 \leq \theta \leq 1 \quad .$$

Then $\|U_\theta(x) - \Phi(x)\| \leq \epsilon/2$ for $x \in \omega$, and for $x \in \partial\omega_N = \partial\omega \cap N$ we have

$$\|U_\theta(x) - x\| \geq \|\Phi(x) - x\| - \|U_\theta(x) - \Phi(x)\| \geq \epsilon - \frac{\epsilon}{2} = \frac{\epsilon}{2} \quad .$$

Therefore, no solutions of $U_\theta(x) - x = 0$ cross $\partial\omega_N$ as θ goes from 0 to 1 , and $d(I - \Phi_\epsilon,\ \omega_N,\ 0) = d(I - \Psi_\epsilon,\ \omega_N,\ 0)$.

Now consider the general case. If we have two approximations Φ_ϵ and Ψ_ϵ on different subspaces M and N , we let L be the subspace containing M and N . Let $\omega_L = \omega \cap L$. The transformations Φ_ϵ and

Ψ_ϵ , being defined everywhere on ω , are defined everywhere on ω_L . By the preceeding argument, $d(I - \Phi_\epsilon,\omega_L,0) = d(I - \Psi_\epsilon,\omega_L,0)$; so it remains to show that $d_L(I - \Phi_\epsilon,\omega_L,0) = d_M(I - \Phi_\epsilon,\omega_M,0)$, and the corresponding result for Ψ_ϵ . That is, we must show that the degree is not changed by enlarging the linear subspace.

We have, essentially, the following situation. A domain ω is situated in $E_n \times E_p$ with $\omega_n = \omega \cap E_n \neq \phi$. Φ is a continuous mapping from ω into E_n . Let x denote points in E_n and y points in E_p . Then the mapping $I - \Phi$ has the form

$$(x,y) \to (x - \varphi(x,y),y) = f(x,y) \tag{7.3.1}$$

where $\varphi : \omega \to E_n$. There are no solutions of $x - \varphi(x,0) = 0$ for $x \in \partial\omega_n$.

We are to show that

$$d_{n+p}(f,\omega,0) = d_n(x - \varphi(x,0),\omega_n,0)$$

where d_{n+p} and d_n denote the degrees in E_{n+p} and E_n respectively. Let

$$f_t(x,y) = (x - t\varphi(x,y) - (1 - t)\varphi(x,0),y)$$

and let ω_t be a homotopy of the domain ω into $\omega_n \times \{-1< y_i < 1\}$. We claim that $d_{n+p}(f_t,\omega_t,0)$ remains constant as t varies. In fact, $f_t = 0$ can have solutions only for $y = 0$; while for $y = 0$, $f_t(x,0) = x - \varphi(x,0)$ which, by assumption, has no zeroes for $x \in \partial\omega_n$. Therefore we have reduced the problem to showing

$$d_{n+p}((x - \varphi(x,0),y),\ \omega_n \times (-1 < y_i < 1),0)$$

$$= d_n(x - \varphi(x,0),\omega_n,0)$$

From now on we write $\varphi(x,0) = \varphi(x)$.

Leray and Schauder base their proof of this point on the definition of degree by simplicial approximations. One can also proceed as follows. Sard's theorem (see J. T. Schwartz, p. 70) states: if f is any C^1 mapping from a domain $\Omega \subset R^n$ then for any measureable subset E of Ω,

$$m(f(E)) \leq \int_E |Jf(x)|dx$$

where $m(f(E))$ is the n dimensional Lebesgue measure of $f(E)$ and $Jf = \det \frac{\partial f_i}{\partial x_j}$. It follows that given any p in $f(\Omega)\setminus f(\partial\Omega)$ there are points q arbitrarily close to p such that $Jf \neq 0$ on the set $f^{-1}(q)$. The preimage set $f^{-1}(q)$ is therefore a discrete set of points, $x_1, \ldots, x_k$, and so

$$d(f,\Omega,p) = d(f,\Omega,q) = \sum_k i[f,x_k] \quad .$$

We apply this observation to the problem at hand, noting first that

$$J(x - \varphi(x),y) = J(x - \varphi(x));$$

and so

$$d_{n+p}((x - \varphi(x),y),\ \omega_n \times (-1,1),0) = d_{n+p}((x - \varphi(x),y),\ \omega_n \times (-1,1),q)$$

$$(q \text{ near } 0)$$

$$= \sum_k i[x_k] = \sum_k \operatorname{sgn} J((x - \varphi,y))\Big|_{\substack{x = x_k \\ y = 0}}$$

$$= \sum_k \operatorname{sgn} J(x - \varphi)\Big|_{x = x_k} = d_n(x - \varphi(x),\omega_n,q_n)$$

$$= d_n(x - \varphi(x),\omega_n,0)$$

(Here q_n denotes the projection of q into E_n).

7.4. Leray's Theorem in hydrodynamics.

In his 1933 thesis Leray considered a number of nonlinear problems, including the boundary value problem for stationary solutions of the Navier-Stokes equations. As Leray pointed out, the method of successive approximations is rather limited in dealing with nonlinear problems. It generally yields only results in the small, and uniqueness of the solutions obtained is difficult to avoid. To get solutions in the large — that is, solutions for all values of a parameter, or without a priori limitations on the smallness of the data of the problem — it is generally necessary to resort to more sophisticated methods.

Leray's thesis was concerned with developing such methods. The development of degree theory by Leray and Schauder followed soon afterward (1934). The methods in the 1934 paper are elegant and simple. It is remarkable that in such a short space of time Leray and Schauder were able to give a complete extension of degree theory to the infinite dimensional case.

In this section we shall apply the degree method to prove an existence theorem in the large for the boundary value problem:

$$\frac{1}{R}\,\Delta u_i - \frac{\partial p}{\partial x_i} = u_j\,\frac{\partial u_i}{\partial x_j}$$

$$\frac{\partial u_i}{\partial x_i} = 0 \qquad u_i\Big|_{\partial D} = \psi_i \tag{7.4.1}$$

The domain D is assumed to be bounded and to have a smooth boundary (say of class $C_{2+\alpha}$). The parameter R is called the Reynolds number. We shall prove that (7.4.1) has a solution for all values of R, provided that

$$\int_{\partial D} \psi_i \nu_i \, ds = 0 \ , \tag{7.4.2}$$

where ν_i are the components of the outward unit normal to the boundary.

Condition (7.4.2) is a necessary consequence of the assumption that $\operatorname{div} u = 0$ in the interior of D. It is a mathematical statement of the fact that the net flux of fluid across the boundary is zero; the condition $\operatorname{div} u = 0$ expresses the condition of incompressibility. We obtain (7.4.2) by integrating the equation $\operatorname{div} u = 0$ over the region D and applying Gauss' theorem.

If (7.4.2) is satisfied and if the ψ_i are smooth functions on ∂D then there is a smooth divergence-free vector field v defined inside D such that $v\big|_{\partial D} = \psi$. (See Ladyzhenskaya, p. 24). Writing $u = w + v$, where u is to be the solution of (7.4.1) we get for w

the system of equations

$$- \frac{1}{R} \Delta w_i + (w_j \frac{\partial v_i}{\partial x_j} + v_j \frac{\partial w_i}{\partial x_j} + w_j \frac{\partial w_i}{\partial x_j}) = - \frac{\partial p}{\partial x_i} + f_i \quad ,$$

$$\frac{\partial w_i}{\partial x_i} = 0 \qquad w_i \Big|_{\partial D} = 0 \quad , \tag{7.4.3}$$

where

$$f_i = -v_j \frac{\partial v_i}{\partial x_j} + \frac{1}{R} \Delta v_i \quad .$$

If ∂D is smooth we can assume that v_i is smooth, and hence that f_i is smooth — say f_i is Hölder continuous. Note that problem (7.4.3) has homogeneous boundary conditions. We must find w_i and p.

We will solve (7.4.3) by converting it to a functional equation in a Hilbert space, as follows. (This procedure is due to Ladyzhenskaya). Define the inner product

$$((u,v)) = \int_D \frac{\partial u_i}{\partial x_j} \frac{\partial v_i}{\partial x_j} dx \qquad \text{(sum on } i \text{ and } j\text{)}$$

and consider the Hilbert space H_σ^1 consisting of all vector fields v in H_σ with finite norm $\|v\| = ((v,v))^{1/2}$.

Let Φ_i be any element of H_σ^1 and multiply (7.4.3) by Φ_i, sum over i, and integrate. After an integration by parts we get

$$\int_D \{\frac{1}{R} \frac{\partial w_i}{\partial x_j} \frac{\partial \Phi_i}{\partial x_j} + w_j \frac{\partial v_i}{\partial x_j} \Phi_i - w_i v_j \frac{\partial \Phi_i}{\partial x_j}$$

$$- w_i w_j \frac{\partial \Phi_i}{\partial x_j}\} dx = \int_D f_i \Phi_i \, dx \tag{7.4.4}$$

We have used the relations

$$w_j \frac{\partial w_i}{\partial x_j} = \frac{\partial}{\partial x_j} w_i w_j \quad \text{and} \quad v_j \frac{\partial w_i}{\partial x_j} = \frac{\partial}{\partial x_j} w_i v_j \quad .$$

We say that w is a generalized, or weak, solution of (7.4.3) if $w \in H^1_\sigma$ and w satisfies (7.4.4) for all $\Phi \in H^1_\sigma$.

Before proceeding we first note the following lemma: (Ladyzhenskaya, Chapter 1, sec. 2).

<u>Lemma 7.4.1. In 3 dimensions, $|u|_4 \leq c\|u\|$ and a weakly convergent sequence of functions in H^1_σ converges strongly in L_4 .</u>

The L_p norm of a vector field u_i is given by

$$|u|_p = \left(\int_D \sum_i |u_i|^p \, dx \right)^{1/p} \quad .$$

A sequence of vector fields w_n is said to converge weakly to w in H^1_σ if $\lim_n((w_n,\Phi)) = ((w,\Phi))$ for every Φ in H^1_σ . The lemma states that if w_n converges weakly to w in H^1_σ then $|w_n - w|_4 \to 0$ as $n \to \infty$.

Now following Ladyzhenskaya, we recast (7.4.4) as a functional equation in H^1_σ . We make use of the Riesz representation theorem: If F is a continuous linear functional on a Hilbert space H with inner product $(\; , \;)$, then there is an element z in H such that $F(x) = (x,z)$ for all $x \in H$.

First consider the functional

$$F(\Phi) = \int_D f_i \Phi_i \, d\underline{x} \quad .$$

Since $|\Phi|_2 \leq c\|\Phi\|$ we have $|F(\Phi)| \leq |f|_2 \cdot |\Phi|_2 \leq c|f|_2\|\Phi\|$. Therefore, F is a bounded linear functional on H^1_σ and by Riesz' theorem there is an element f in H^1_σ such that $F(\Phi) = ((f,\Phi))$. (We are working on a real Hilbert space so $((f,\Phi)) = ((\Phi,f))$.)

The operation

$$F(\Phi) = \int_D \frac{\partial w_i}{\partial x_j} \frac{\partial \Phi_i}{\partial x_j} \, d\underline{x}$$

is simply the functional $F(\Phi) = ((w,\Phi))$. The operation

$$F(\Phi) = \int_D u_i w_j \frac{\partial \Phi_i}{\partial x_j} \, d\underline{x}$$

is a bounded linear functional on H^1_σ if $u,w \in H^1_\sigma$. In fact, by Hölder's inequality we have

$$|F(\Phi)| \leq \left(\int_D \sum_{i,j} (u_i w_j)^2 \, dx \right)^{1/2} \left(\int_D \frac{\partial \Phi_i}{\partial x_j} \frac{\partial \Phi_i}{\partial x_j} \, dx \right)^{1/2}$$

$$= \left(\int_D (\sum_i u_i^2)(\sum_j w_j^2) \, dx \right)^{1/2} \|\Phi\| \qquad (7.4.5)$$

$$\leq c|u|_4 |w|_4 \|\Phi\| \leq c'\|u\| \cdot \|w\| \cdot \|\Phi\|$$

Therefore there is an element, say $N(u,w)$, in H^1_σ such that

$$((N(u,w),\Phi)) = \int_D u_i w_j \frac{\partial \Phi_i}{\partial x_j} \, dx \quad .$$

Clearly N is a bilinear operator from $H^1_\sigma \times H^1_\sigma$ to H^1_σ .

Similarly, there is an operator L from H^1_σ to itself defined by

$$((Lw,\Phi)) = \int_D \{w_j \frac{\partial v_i}{\partial x_j} \Phi_i - v_i w_i \frac{\partial \Phi_i}{\partial x_j}\} dx .$$

Thus we may write (7.4.4) in the form

$$((\frac{1}{R} w + Lw - N(w,w) - F, \Phi)) = 0 \qquad (7.4.6)$$

This equation is to hold for all $\Phi \in H^1_\sigma$, so we must have

$$w + R(Lw - N(w,w) - F) = 0 \qquad (7.4.7)$$

We are going to show

(i) L and N are completely continuous on H^1_σ.

(ii) Solutions of (7.4.7) are a priori bounded: $\|w\| \leq R(a + b\|F\|^2)^{1/2}$.

Then we shall apply a degree of mapping argument to show that (7.4.7) has a solution for all values of R.

First let us show that N is completely continuous. Suppose w_n is a weakly convergent sequence in H^1_σ. We have, by 7.4.5,

$$|((N(w_n,w_n),\Phi)) - ((N(w_m,w_m),\Phi))|$$

$$= |((N(w_n - w_m,w_n) - N(w_m,w_m - w_n),\Phi))|$$

$$\leq c|w_n - w_m|_4(|w_n|_4 + |w_m|_4) \|\Phi\| .$$

Since this holds for all Φ, we may take $\Phi = N(w_n,w_n) - N(w_m,w_m)$, obtaining

$$\|N(w_n,w_n) - N(w_m,w_m)\| \leq c|w_n - w_m|_4(|w_n|_4 + |w_m|_4) .$$

Since $|w_n - w_m|_4 \to 0$ as $n,m \to \infty$, $|w_n|_4$ and $|w_m|_4$ remain uniformly bounded as $n,m \to \infty$, and so $\{N(w_n,w_n)\}$ is a Cauchy sequence in H^1_σ. The complete continuity of L may be established in a similar manner.

Let us now derive the a priori bounds. Taking the inner product of (7.4.7) with w we get

$$\|w\|^2 + R\{((Lw,w)) + ((N(w,w),w)) - ((F,w))\} = 0 . \qquad (7.4.8)$$

Now

$$((N(w,w),w)) = \int_D w_i w_j \frac{\partial w_i}{\partial x_j} dx$$

$$= \int_D w_j \frac{\partial}{\partial x_j} \frac{w^2}{2} dx = 0$$

(This argument certainly holds if w is smooth and vanishes on ∂D and can be extended to all vector fields in H^1_σ by continuity.) Similarly,

$$((Lw,w)) = \int_D \{w_i w_j \frac{\partial v_i}{\partial x_j} - v_j w_i \frac{\partial w_i}{\partial x_j}\} dx$$

$$= \int_D w_i w_j \frac{\partial v_i}{\partial x_j} d\underline{x} .$$

We estimate this term following a procedure due to E. Hopf [3]. Hopf proved that for any $\epsilon > 0$ the vector field v can be so chosen that

$$|\int_D w_i w_j \frac{\partial v_i}{\partial x_j} dx| \le \epsilon \|w\|^2 \qquad (7.4.9)$$

We shall prove Hopf's lemma below. Meanwhile, we first complete the proof of Leray's theorem. By virtue of (7.4.8) and (7.4.9),

$$\|w\|^2 \leq \epsilon R\|w\|^2 + R\|F\|\cdot\|w\| ;$$

hence, choosing ϵ and v so that $\epsilon R < 1$,

$$\|w\| \leq \frac{R}{1 - \epsilon R} \|F\| . \tag{7.4.10}$$

With (i) and (ii) established we can prove Leray's existence theorem. Let $R_0 > 0$, and choose ϵ and v so that $\epsilon R_0 < 1$. Consider the ball B in H^1_σ of radius $\dfrac{1 + R_0}{1 - \epsilon R_0}$. For $0 < R < R_0$ any solution of (7.4.7) must be bounded in norm by $\dfrac{R}{1 - \epsilon R} < \dfrac{1 + R_0}{1 - \epsilon R_0}$, so no solutions can cross the boundary of B. Thus the degree

$$d(I + R(L - N - F), \Omega, 0)$$

remains constant for $0 \leq R < R_0$. For $R = 0$ the degree is 1 so it must remain 1. Therefore (7.4.7) has a solution for all $R < R_0$. Since R_0 was arbitrary, (7.4.7) has a solution for all R.

It remains to prove Hopf's lemma. For $x \in D$ let $s(x)$ denote the distance from x to ∂D (which is assumed to be smooth). For $p > 0$ let $D_p = \{x : x \in D, s(x) \leq p\}$. For sufficiently small p, D_p has a smooth boundary. From the Hardy-Littlewood inequality

$$\int_0^p \left(\frac{f(s)}{s}\right)^2 ds \leq 4 \int_0^p (f'(s))^2 ds ,$$

where $f \in C^1$ and $f(0) = 0$, it follows that

$$\int_{D_p} \left(\frac{f(x)}{s(x)} \right)^2 dx \leq C \int_{D_p} \frac{\partial f}{\partial x_j} \frac{\partial f}{\partial x_j} dx$$

for every f in $C^1(D)$ which vanishes on ∂D. We shall construct a divergence free vector field v in D such that

$$\left| \frac{\partial v_i}{\partial x_j} \right| < \frac{\epsilon}{s^2(x)} , \tag{7.4.11}$$

v satisfies the required boundary conditions on ∂D, and $v \equiv 0$ in $D - D_p$. Then

$$\left| \int_D w_i w_j \frac{\partial v_i}{\partial x_j} dx \right| \leq \epsilon \left| \int_{D_p} \frac{w_i w_i}{s^2(x)} dx \right|$$

$$\leq C \epsilon \int_{D_p} \frac{\partial w_i}{\partial x_j} \frac{\partial w_i}{\partial x_j} dx \leq C \epsilon \|w\|^2 .$$

First, as shown in Ladyzhenskaya, one can construct a vector field b such that $a = \text{curl}\, b$ satisfies the given boundary conditions on ∂D, while a is automatically divergence free. We look for v in the form

$$v = \text{curl}\, h(s(x))b(x)$$

where $h(s)$ is an appropriately chosen scalar valued function. In particular, we want $h \in C^3$, $h(0) = 1$, $h'(0) = 0$, and $h = 0$ for $s \geq p$. It is then clear that $\text{div}\, v = 0$, $v = 0$ in $D - D_p$, and v satisfies the correct boundary conditions on ∂D. In order that (7.4.11) hold we must require in addition that the second derivatives of $h(s(x)b(x)$ are bounded in absolute value by $\epsilon\, s^{-2}(x)$. Since b_i and $s(x)$ are fixed, it is sufficient to construct h so that

$$|h(s)| + |h'(s)| + h''(s)| < \frac{\epsilon}{s^2}$$

This is easily done as follows. Choose $k \in C^3$ such that $k(0) = 1$, $k'(0) = 0$ and $k = 0$ for $s \geq p$. Then given any $\epsilon > 0$ there is a $\delta > 0$ such that the function

$$h(s) = \frac{\int_\delta^1 \frac{1}{t}\, k(\frac{s}{t})\, dt}{\log \frac{1}{\delta}}$$

satisfies the required conditions. The verification of this fact is left to the reader.

<u>Exercises</u>:

1. Let $f : R^2 \to R^2$ be defined by $f(x,y) = (x^2 + y^2 -1, 0)$. Let $\Omega = \{(x,y) \mid x^2 + y^2 \leq 2\}$ and compute $d(f,\Omega,0)$.

2. Try to prove an existence theorem for (7.4.1) for all values of R by successive approximations and by the continuity method.

3. Using Hopf's lemma, derive *a priori* estimates for the time dependent Navier-Stokes equations. Show in particular that $|u(t)|$ is uniformly bounded for all $t > 0$.

4. Can you apply a degree argument to prove an existence theorem for the time dependent Navier-Stokes equations? (Use the *a priori* estimates of Ex. 3)

5. Try to apply a degree argument to problems of the type

$$\Delta u + f(x,u,\nabla u) = 0 .$$

6. Prove that if $w \in H^1_\sigma$ satisfies (7.4.4) and if w is C^2, then w satisfies (7.4.3). (Hint: use the fact that $L_2 = H_\sigma \oplus H_\pi$, where H_π is the subspace of gradients. (See Ladyzhenskaya, p. 27)

7.5. Bifurcation by Leray Schauder degree.

In this section we discuss the use of topological degree arguments to establish the existence of bifurcating solutions. We consider a functional equation

$$(I - \mu T)u + F(\mu,u) = 0 \tag{7.5.1}$$

on a Banach space $\mathcal{B}$, where T and F are completely continuous. We assume that F is uniformly continuous in μ as u varies over bounded sets and that $F(\mu,0) \equiv 0$. Let $E = R \times \mathcal{B}$, where R is the set of real numbers; E is a Banach space with norm $\|(\mu,u)\| = \sqrt{|\mu|^2 + \|u\|^2}$. A solution of (7.5.1) is pair (μ,u) belonging to E.

Lemma 7.5.1. Assume that for $\mu = \mu_0$ there exists an $\epsilon > 0$ such that (7.5.1) has no solutions of norm ϵ. Then there is a $\delta > 0$ such that (7.5.1) has no solutions of norm ϵ for $|\mu - \mu_0| < \delta$.

Proof: Suppose the conclusion does not hold. Then there is a sequence (μ_n,u_n) of solutions of (7.5.1) such that $\|u_n\| = \epsilon$ and μ_n tends to μ_0. Since T and F are completely continuous, and since F is uniformly continuous in μ, there is a subsequence (which we again denote by (μ_n,u_n)) such that $F(\mu_n,u_n)$ and Tu_n

are convergent. It follows from (7.5.1) that u_n converges as well. Taking the limit as $n \to \infty$ we get a solution (μ_0, u) of (7.5.1) for which $\|u\| = \epsilon$. This contradicts the hypothesis of the lemma.

<u>Theorem 7.5.2. Let μ_0 be a characteristic value of T of odd multiplicity and let F be Frechet differentiable at $u = 0$ with vanishing derivative near $\mu = \mu_0$. Then (7.5.1) has nontrivial solutions in a suitably small rectangle $\{\|u\| < \epsilon, |\mu - \mu_0| < \delta\}$.</u>

<u>Proof</u>: If the hypothesis of Lemma (7.5.1) is not satisfied then the conclusion of Theorem 7.5.2 obviously holds. This is the situation, for example, if $F \equiv 0$ (the linear case) ; the case of vertical bifurcation may be considered singular. Therefore, consider the case where the hypothesis of Lemma (7.5.1) is valid, and let ϵ and δ be as given there. Let $\Omega = \{u : \|u\| < \epsilon\}$ and consider

$$\deg((I - \mu T) + F(\mu, \cdot), \Omega, 0) .$$

If $u = 0$ is the only solution of (7.5.1) then the degree is the index of $u = 0$. But since $F'_u(\mu, 0) = 0$,

$$i[0] = \deg(I - \mu T, \Omega, 0) .$$

According to property (vii) of the Leray Schauder degree, this index changes by a factor of -1 as μ crosses μ_0. On the other hand, the degree must remain constant as μ varies over the interval $(\mu_0 - \delta, \mu_0 + \delta)$, where δ is given by Lemma 7.5.1. Therefore $u = 0$ cannot be the only solution of (7.5.1) in the rectangle $\{\|u\| < \epsilon, |\mu - \mu_0| < \delta\}$.

A corollary of the above result is the following: _If (7.5.1) has no nontrivial solutions for $\mu_0 - \delta < \mu \le \mu_0$, then it must have nontrivial solutions for $\mu_0 < \mu < \mu_0 + \delta$._

Judovic [6] and Velte applied the above argument to prove the existence of non-trivial solutions in convection problems for the Navier-Stokes Equations. They reduced the relevant systems of partial differential equations to a functional equation of type (7.5.1) and proved that for $\mu < \mu_0$ the only solution is the null solution.

Theorem (7.5.2) is a local result, but Rabinowitz [2] has shown that bifurcation from an eigenvalue of odd multiplicity is a global rather than a local phenomenon. A _continuum_ of solutions of (7.5.1) is defined to be a closed connected set $\{(\mu,u)\}$ of points in E satisfying (7.5.1). Let $\mathfrak{S}$ denote the set of non-trivial solutions of (7.5.1).

Theorem 7.5.3. (Rabinowitz) _If μ is a characteristic value of T of odd multiplicity then there is a maximal continuum of solutions C_μ containing nontrivial solutions such that $(\mu,0) \in C_\mu$ and C_μ either (i) tends to infinity in E or (ii) meets a point $(\mu_1,0)$ where μ_1 is another characteristic value of T. If μ is a simple characteristic value of T then near $(\mu,0)$ $C_\mu \setminus \{(\mu,0)\}$ consists of two distinct subcontinua which meet only at $(\mu,0)$._

Rabinowitz's methods do not require that F be Frechet differentiable. One interesting application of theorem 7.5.3 is the following. Consider the nonlinear Sturm-Liouville problem

$$\mathcal{L}u = -(pu') + qu = F(x,u,u',\mu) \qquad 0 \leq x \leq \pi$$

$$a_0 u(0) + b_0 u'(0) = 0 \tag{7.5.2}$$

$$a_1 u(\pi) + b_1 u'(\pi) = 0$$

where $(a_0^2 + b_0^2)(a_1^2 + b_1^2) \neq 0$. The function F has the form

$$F(x,u,u',\mu) = \mu a(x)u + H(x,u,u',\mu)$$

where $H = 0(\sqrt{u^2 + u'^2})$ near $u = u' = 0$. The bifurcation points of (7.5.2) are $u = 0$ and $\mu = \mu_n$, where μ_n is an eigenvalue of $\mathcal{L}$ with the given boundary conditions: that is $\mathcal{L}\varphi - \mu_n a\varphi = 0$.

The eigenvalues of $\mathcal{L}$ are all real and simple and tend to $\infty : \mu_1 < \mu_2 < \ldots$. The nth eigenfunction φ_n has precisely $(n-1)$ simple zeroes in $(0,\pi)$.

If $\mathcal{L}$ is invertible (7.5.2) can be converted into an integral equation using Green's function. One then obtains a functional equation of the type (7.5.1). Define the set S_n^+ to consist of functions in $C^1[0,\pi]$ which have precisely $(n-1)$ simple zeroes and which are positive for x near 0 . The set S_n^- consists of those functions u such that $-u$ belongs to S_n^+ . From each eigenvalue μ_n there bifurcate two continua of solutions according to Theorem (7.5.3). It can be shown that one continuum lies in S_n^+ while the other lies in S_n^- . Furthermore, any solution of (7.5.2) on the boundary of S_n^+ or S_n^- must vanish identically. In fact, the boundaries of S_n^+ and S_n^- consist of solutions of (7.5.2) with a double zero and such solutions vanish identically by the uniqueness theorem for the ordinary differential equations (7.5.2).

Therefore a continuum from $(\mu_n, 0)$ cannot go over to $(\mu_m, 0)$ $(\mu_m \neq \mu_n)$ because the respective continua have different nodal properties. Accordingly, alternative (i) of theorem (7.5.3) must hold.

Pictorially, we might represent the picture like this:

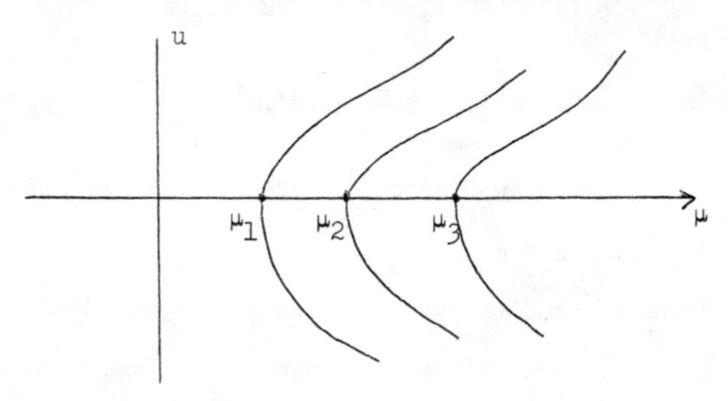

These results for the Sturm-Liouville problem (7.5.2) were originally obtained by Crandall and Rabinowitz [2] by calculating the degree of the functional equation relative to the sets S_n^+ and S_n^-.

7.6. A Result of H. Amann.

In Chapter 2 we considered boundary value problems of the form

$$Lu + f(x,u) = 0 \qquad u\big|_{\partial D} = 0 . \tag{7.6.1}$$

where L is a second order elliptic operator. We proved (theorem 2.3.1) that if ψ is a lower solution and ϕ is an upper solution, with $\psi < \phi$, then there is at least one solution of (7.6.1) lying between ψ and ϕ. In fact, referring to the proof of that theorem we see that we get a minimal solution $\underline{u}$ and a maximal solution $\bar{u}$.

The minimal solution is obtained as the limit of a sequence increasing from ψ , while $\bar{u}$ is the limit of a sequence decreasing from ϕ .

It may be that $\underline{u} \neq \bar{u}$. In that case, H. Amann has proved the following result using an interesting degree argument.

Theorem 7.6.1 (Amann): Let the hypotheses of Theorem (2.3.1) be satisfied and suppose there exist two distinct solutions of (7.6.1), $u_1 < u_2$, such that $\psi < u_1 < u_2 < \phi$. Moreover, assume that the derived operators

$$L + f'_u(x,u) , \quad i = 1,2$$

with homogeneous boundary conditions are invertible. Then (7.6.1) has at least one more distinct solution u_3, $u_1 < u_3 < u_2$.

Using Green's function for the operator L, (7.6.1) can be written in the form

$$u + Kf(\cdot,u) = 0 \tag{7.6.2}$$

where K is the integral operator

$$Kv = \int_D K(x,y)v(y)dy .$$

Using the a priori estimate

$$\|u\|_{W^2_p} \leq c\|Lu\|_{L_p}$$

of Agmon-Douglis-Nirenberg we can show that K is a completely continuous transformation on the Banach space $C(\bar{D})$ (continuous functions on $\bar{D}$ with the norm $\|u\| = \sup_{x \in \bar{D}} |u(x)|$). The L_p

estimates imply that if $Lu = f$ and $f \in L_p$ then $u \in W_p^2$ (the Sobolev space of functions with weak second derivatives in L_p). If f is in L_∞ then it is in L_p for all $p \geq 1$, and so $u \in W_p^2$ for all $p \geq 1$. But for $p > n$, $W_p^2(D) \subset C^{1+\alpha}(\bar{D})$ for some α, $0 < \alpha < 1$ (see Chapter 2). Since bounded sets in $C^{1+\alpha}(\bar{D})$ are compact in $C(\bar{D})$ we have established the complete continuity of the integral operator K.

Lemma 7.6.2: Let A be a completely continuous transformation on a Banach space E. Suppose B is a linear compact operator such that $(I + B)$ is invertible and

$$\lim_{\|u\| \to \infty} \frac{\|A(u) - Bu\|}{\|u\|} = 0 \tag{7.6.3}$$

Let $u_i + A(u_i) = 0$, $i = 1,2,$ and suppose $I + A'(u_i)$ is invertible, $i = 1,2,$. Then there is at least one other solution of $u + A(u) = 0$.

Proof: Since $I + B$ is invertible there exists a constant a such that

$$\|u + Bu\| \geq a\|u\| .$$

From (7.6.3) there exists R such that for all u with norm greater than R,

$$\|A(u) - Bu\| \leq \frac{a}{2} \|u\| .$$

Therefore, for $\|u\| \geq R$ and $0 \leq \tau \leq 1$,

$$\|u + \tau A(u) + (1 - \tau)Bu\| \geq \|u + Bu\| - \|A(u) - Bu\| \geq \frac{aR}{2} .$$

This shows that in the ball $B_R = \{u : \|u\| < R\}$,

$$\deg(I + A, B_R, 0) = \deg(I + B, B_R, 0) \quad .$$

Since $(I + B)$ is invertible, u_1 and u_2 are isolated solutions, and their indices are ± 1 . Recall that $\deg(I + A, B_R, 0)$ is the sum of the indices of the solutions of $u + A(u) = 0$. If u_1 and u_2 were the only solutions, this degree would be -2, 0, or 2, so there must be at least one other solution. More generally, we have shown that the degree of $I + A$ in a sufficiently large ball is ± 1 .

Now let us prove Theorem (7.6.1). Define

$$g(x,\xi) = \begin{cases} f(x,u_1(x)) + u_1(x) - \xi & \text{if } \xi \le u_1(x) \\ f(x,\xi) & \text{if } u_1(x) \le \xi \le u_2(x) \\ f(x,u_2(x)) + u_2(x) - \xi & \text{if } \xi \ge u_2(x) \end{cases}$$

If f is Lipshitz continuous in x and u then so is g . Let us show that every solution of the boundary value problem

$$Lu + g(x,u) = 0 \ , \quad u\Big|_{\partial D} = 0 \tag{7.6.4}$$

is also a solution of (7.6.1). Let u be a solution of (7.6.4) and let $D_1 = \{x \in D : u(x) < u_1(x)\}$. Then $w = u - u_1$ satisfies

$$Lw = Lu - Lu_1 = -g(x,u) + f(x,u_1) = u - u_1 \le 0 \quad \text{in } D_1$$

By the strong maximum principle, w cannot attain a minimum in the interior of D_1 unless w is identically constant there. Therefore

D_1 is empty (since $w = 0$ on ∂D_1) and $u(x) \geq u_1(x)$ everywhere. Similarly, $u \leq u_2$ everywhere, so the two boundary problems are identical.

Thus it is sufficient to show that (7.6.4) has three solutions. Operating on (7.6.4) with $L^{-1} = K$ we have

$$u + Kg(x,u) = 0 \ .$$

Now $g(x,\xi) + \xi$ is uniformly bounded for all ξ, so

$$\lim_{\|u\| \to \infty} \frac{\|Kg(x,u) + Ku\|}{\|u\|} \leq \lim_{\|u\| \to \infty} \frac{\|K\| \ \|g(x,u) + u\|}{\|u\|} = 0 \ .$$

Furthermore $(I - K)$ is invertible. (By the maximum principle $L\varphi - \varphi = 0$ cannot have nontrivial solutions). So we may apply lemma (7.6.2) with $B = -K$, and the proof of Theorem (7.6.1) follows immediately.

VIII

The Real World

In the previous chapters we have discussed mathematical techniques for investigating the stability and bifurcation of solutions of the Navier Stokes equations. In this chapter we shall discuss briefly some specific physical problems for which there is an abundant experimental as well as theoretical literature.

In much of the physical literature the analysis is carried out on an unbounded domain with spatial periodicity imposed upon the functions. The problem of carrying out bifurcation analyses in a class of periodic functions generally presents no formal difficulties; the only modification really necessary is that of obtaining *a priori* estimates of Solonnikov's type for spatially periodic functions instead of functions in a bounded domain.

To illustrate the role that periodicity plays in stability investigations, let us consider the problem of flow through an infinitely long pipe

$$(x,y) \in \Omega \,, \quad -\infty < z < \infty \,,$$

where Ω is a bounded domain in the x-y plane. Suppose that the steady flow has the form

$$\tilde{u}(x,y,z) = (0,0,\ w(x,y)) \quad .$$

Then the linearized eigenvalue problem to be considered would be

$$\frac{1}{R}\Delta\varphi_1 + w\frac{\partial\varphi_1}{\partial z} = -\frac{\partial p}{\partial x} + \lambda\varphi_1$$

$$\frac{1}{R}\Delta\varphi_2 + w\frac{\partial\varphi_2}{\partial z} = -\frac{\partial p}{\partial z} + \lambda\varphi_2$$

$$\frac{1}{R}\Delta\varphi_3 + w\frac{\partial\varphi_3}{\partial z} + w_x\varphi_1 + w_y\varphi_2 = -\frac{\partial p}{\partial z} + \lambda\varphi_3$$

$$\frac{\partial\varphi_1}{\partial x} + \frac{\partial\varphi_2}{\partial y} + \frac{\partial\varphi_3}{\partial z} = 0 \qquad \varphi_i\Big|_{\partial\Omega} = 0$$

This eigenvalue problem would be investigated by separating variables and writing

$$\varphi_k(x,y,z) = e^{i\alpha z}\,\psi_k(x,y) \qquad k = 1,2,3$$

$$p(x,y,z) = e^{i\alpha z}\,q(x,y)$$

This leads to an eigenvalue problem for ψ on Ω in which there appear two parameters R and α. The number α is called the wave number.

For stability of the basic flow $\tilde{u}$ we should require that it be stable with respect to disturbances of arbitrary wave number. For each wave number α there is a Reynolds number R at which the flow loses stability. The critical Reynolds number would be the smallest value of R for which the flow loses stability with respect to disturbances of some given wave number.

By this procedure a critical wave number is obtained as well as a critical Reynolds number. See, for example, the book of C. C. Lin. The bifurcation analysis is then carried out in the class of functions of the critical wave number.

With this in mind, we now pass to a brief discussion of three physical problems — the Benard problem, the Taylor problem, and Poiseuille flow in pipes.

8.1. The Benard Problem.

Consider a layer of fluid infinite in horizontal extent and lying in the region $0 \leq x_3 \leq d$. The fluid density is assumed to vary with temperature according to the law

$$\rho = 1 - \alpha(T - T_0) \quad .$$

In the simplest model, the Boussinesq approximation, variations in density are neglected except in the buoyant force term. The equations of motion then take the form

$$\frac{\partial u_i}{\partial t} + u_j \frac{\partial u_i}{\partial x_j} = - \frac{\partial p}{\partial x_i} - \rho g \delta_{i3} + \nu \Delta u_i \quad ,$$

$$\frac{\partial T}{\partial t} + u_j \frac{\partial T}{\partial x_j} = k \, \Delta \, T \quad , \tag{8.1.1}$$

$$\frac{\partial u_i}{\partial x_i} = 0 \quad ,$$

$$T(x_1,x_2,0) = T_0 \ , \quad T(x_1,x_2,d) = T_1 \qquad u_i = 0 \ \text{ at } \ x_3 = 0,d$$

In these equations, δ_{i3} is Kronecker's function and g is the acceleration due to gravity.

When $T_0 > T_1$ the buoyant force term $-\rho g \delta_{i3}$ acts to destabilize the fluid. Beyond a critical temperature drop across the layer convection sets in — that is, a steady state solution of (8.1.1) exists for which the velocity field u assumes non-zero, time independent values.

In the pure conduction state we have $u = 0$ and the steady state equations (8.1.1) take the form

$$p_{x_1} = p_{x_2} = 0$$

$$p_{x_3} = -(1 - \alpha(T - T_0))g$$

$$\tilde{T}_{x_3 x_3} = 0$$

The solution for $\tilde{T}$ is

$$\tilde{T}(x_3) = T_0 + (T_1 - T_0)\frac{x_3}{d}$$

Denoting the perturbation from steady state by θ, $\theta = T - \tilde{T}$, we get the following equations for the perturbations:

$$\frac{\partial u_i}{\partial t} + u_j \frac{\partial u_i}{\partial x_j} = -\frac{\partial p}{\partial x_i} + \nu \Delta u_i + \alpha g \theta \delta_{i3}$$

$$\frac{\partial \theta}{\partial t} + u_j \frac{\partial \theta}{\partial x_j} = k \Delta \theta - \frac{(T_1 - T_0)}{d} u_3$$

$$\frac{\partial u_i}{\partial x_i} = 0$$

We now make the following scaling changes

$$\xi_i = \frac{x_i}{d} \qquad \tau = \frac{k}{d^2}\, t \qquad u_4 = \frac{d^2 \alpha g}{\nu}\, \theta\,, \qquad q = \frac{d}{k} p\ .$$

The equations then take the following form

$$\frac{\partial u_i}{\partial \tau} - P(\Delta u_i + u_4 \delta_{i3}) = -\frac{\partial q}{\partial \xi_i} - \frac{d}{k}\, u_j\, \frac{\partial u_i}{\partial \xi_j}$$

$$\frac{\partial u_4}{\partial \tau} - \Delta u_4 + R u_3 = -\frac{d}{k}\, u_j\, \frac{\partial u_4}{\partial \xi_j} \tag{8.1.2}$$

$$\frac{\partial u_i}{\partial x_i} = 0$$

where $P = \nu/k$ is the Prandtl number and $R = (T_1 - T_0)d^3 \alpha g/k\nu$ is the Rayleigh number. Summation is only over $i = 1,2,3$. In an experimental situation, P remains constant while R measures the temperature drop across the layer.

Equations (8.1.2) are in many respects similar to the Navier Stokes equations themselves, and can be treated by the same methods as in Chapters IV through VII.

Equations (8.1.2) are analyzed for functions periodic in ξ_1 and ξ_2. The three possibilities are rectangular, and hexagonal cellular patterns and "rolls" (solutions independent of ξ_2 and periodic in ξ_1). Formal investigations of the bifurcating convective solutions were carried out by Malkus and Veronis. Rigorous demonstrations of the bifurcation were given by Judovic [6] and, for a related problem, by Velte, using topological degree arguments. Rabinowitz [1] then gave a constructive proof. Recent investigations by Joseph [2], Fife and Joseph, and Fife [1] have carried out a deeper analysis by considering convection problems governed by more general systems of equations. These models allow a wider variety of convection effects, for example the phenomenon of subcritical bifurcation. Fife considers very general fluid dynamical equations in which the stress tensor is allowed to depend nonlinearly on the temperature and strain tensor; he treats the bifurcation problem rigorously and establishes the validity of Boussinesq approximations.

Joseph [2] discusses the phenomenon of subcritical bifurcation general convection problems. He shows that under certain circumstances the bifurcation diagram takes the form shown below.

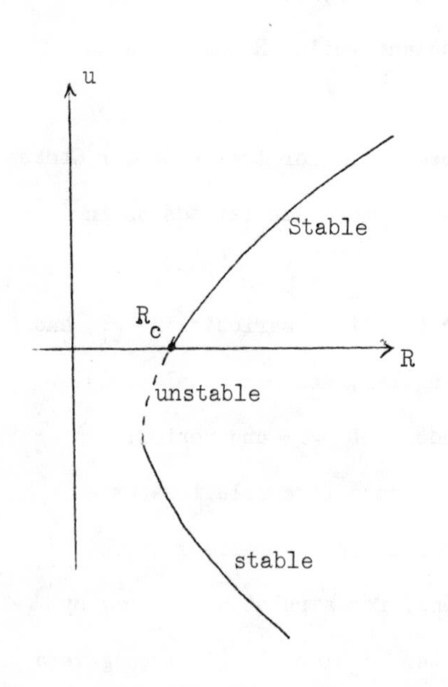

The conduction solution lies along the R-axis. At $(R_c, 0)$ we get a bifurcating curve, with one branch supercritical and the other subcritical. The subcritical branch "turns back" and regains stability after the turn. If the fluid moves to the stable section of the subcritical branch, the Rayleigh number can be reduced below R_c before the convection disappears. Also, below criticality, a sufficiently large displacement from equilibrium might cause the system to pass through the unstable convective solution and come to rest on the mode of stable, subcritical convection.

For further discussion of convention problems, see the article by L. A. Segel.

8.2 Flow between rotating cylinders.

We consider the flow of fluid between concentric cylinders of radii R_1 and R_2 which rotate with angular velocities Ω_1 and Ω_2 respectively. The Navier Stokes equations in cylindrical coordinates are

$$\frac{\partial u}{\partial t} + u\,\frac{\partial u}{\partial r} + \frac{v}{r}\,\frac{\partial u}{\partial \theta} + w\,\frac{\partial u}{\partial z} - \frac{v^2}{r}$$

$$= - \frac{\partial p}{\partial r} + \frac{1}{R}\left(\Delta u - \frac{u}{r^2} - \frac{2}{r^2}\,\frac{\partial v}{\partial \theta}\right)$$

$$\frac{\partial v}{\partial t} + u\,\frac{\partial v}{\partial r} + \frac{v}{r}\,\frac{\partial v}{\partial \theta} + w\,\frac{\partial v}{\partial z} + \frac{uv}{r} \qquad (8.2.1)$$

$$= - \frac{1}{r}\,\frac{\partial p}{\partial \theta} + \frac{1}{R}\left(\Delta v - \frac{v}{r^2} + \frac{2}{r^2}\,\frac{\partial u}{\partial \theta}\right)$$

$$\frac{\partial w}{\partial t} + u\,\frac{\partial w}{\partial r} + \frac{v}{r}\,\frac{\partial w}{\partial \theta} + w\,\frac{\partial w}{\partial z}$$

$$= - \frac{\partial p}{\partial z} + \frac{1}{R}\,\Delta\, w$$

$$\frac{1}{r}\,\frac{\partial}{\partial r}(ru) + \frac{1}{r}\,\frac{\partial v}{\partial \theta} + \frac{\partial w}{\partial z} = 0$$

These equations admit a stationary solution of the form

$$u = w = 0\,, \qquad p = p(r)\,, \qquad v = \tilde{v}(r)$$

where

$$\tilde{v}(r) = Ar + \frac{B}{r}\;.$$

The angular velocity is then $\tilde{\Omega} = \tilde{U}/r = A + B/r^2$. Applying the boundary conditions we get

$$A = -\Omega_1 \eta^2 \frac{1 - \mu/\eta^2}{1 - \eta^2}$$

$$B = \Omega_1 \frac{R_1^2 (1 - \mu)}{1 - \eta^2}$$

where

$$\mu = \frac{\Omega_2}{\Omega_1} , \qquad \eta = \frac{R_1}{R_2}$$

The eigenvalue problem which determines the stability of this basic flow is

$$\lambda u + \frac{\tilde{v}}{r}\frac{\partial u}{\partial \theta} - \frac{2\tilde{v}v}{r} = -\frac{\partial p}{\partial r} + \frac{1}{R}\left(\Delta u - \frac{u}{r^2} - \frac{2}{r^2}\frac{\partial v}{\partial \theta}\right)$$

$$\lambda v + 2Au + \frac{\tilde{v}}{r}\frac{\partial v}{\partial \theta} = -\frac{1}{r}\frac{\partial p}{\partial \theta} + \frac{1}{R}\left(\Delta v - \frac{v}{r^2} + \frac{2}{r^2}\frac{\partial u}{\partial \theta}\right)$$

$$\lambda w + \frac{\tilde{v}}{r}\frac{\partial w}{\partial \theta} = -\frac{\partial p}{\partial z} + \frac{1}{R}\Delta w$$

$$\frac{1}{r}\frac{\partial}{\partial r}(ru) + \frac{1}{r}\frac{\partial v}{\partial \theta} + \frac{\partial w}{\partial z} = 0$$

These equations admit the following separation of variables:

$$u = \cos \alpha z \; e^{im\theta} U(r)$$

$$v = \cos \alpha z \; e^{im\theta} V(r)$$

$$w = \sin \alpha z \; e^{im\theta} W(r)$$

$$p = \cos \alpha z \; e^{im\theta} P(r)$$

We then obtain an eigenvalue problem for U , V , W , P on $R_1 \leq r \leq R_2$. The general eigenvalue problem is quite complex, and a complete analysis is practically impossible. In practice a number of simplifying assumptions are made. For example, experimental evidence indicates that in some situations the first mode to go unstable is axisymmetric. In that case one may take $m = 0$ above and obtain the following eigenvalue problem:

$$\lambda U - \frac{2v}{r} V = -P^1 + \frac{1}{R}\left(\frac{1}{r}(rU')' - \frac{m^2 + \alpha^2 + 1}{r^2} U\right)$$

$$\lambda V + 2AU = \frac{1}{R}\left(\frac{1}{r}(rV')' - \frac{m^2 + \alpha^2 + 1}{r^2} V\right)$$

$$\lambda W = -\alpha P + \frac{1}{R}\left(\frac{1}{r}(rW')' - \frac{m^2 + \alpha^2}{r^2} W\right)$$

$$\frac{1}{r}(rU)' + \alpha W = 0$$

$$U = V = W = 0 \quad \text{on} \quad r = R_1 , R_2 .$$

An extensive analysis of this eigenvalue problem is carried out in Chandrasekhar's book. Within certain ranges of R , η , and μ it is known that the flow loses stability by virtue of a real eigenvalue crossing the origin. A critical wave number α_0 is obtained, relative to which the flow first loses stability. The bifurcation analysis is carried out in the class of flows periodic in the z direction with wave number α_0 . (See the article by Kirchgässner and Sorger.) The disturbances which appear are called Taylor vortices, after G. I. Taylor who in 1923 first gave an accurate theoretical and experimental account of the phenomenon. Some of Taylor's original photographs may be found in Chandrasekhar's book .

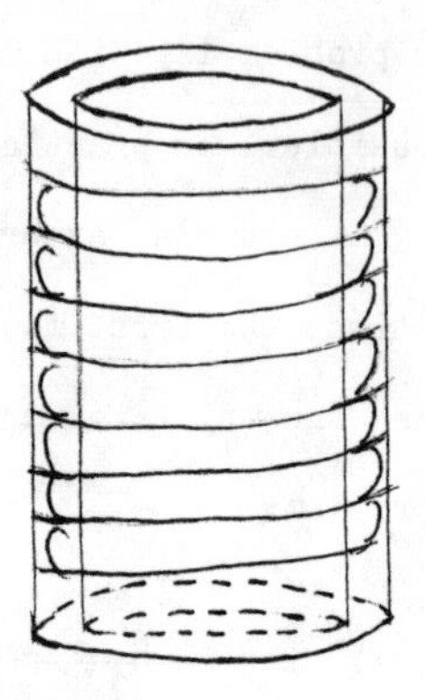

The first point at which the flow loses stability may thus result in the bifurcation of a new stationary flow. When the relative rate of rotation is increased still further, the Taylor vortices lose their stability and are replaced by flows of more complicated behavior. In some cases, these may take the form of oscillations (traveling waves) (see D. Coles and Davey, di Prima, and Stuart). A discussion of the bifurcation of Taylor vortices may be found in the two articles by Kirchgässner and Sorger [1,2]; also, see Kirchgässner [1] .

8.3. Plane Poiseuille Flow.

A special solution of the Navier Stokes equations is that of plane Poiseuille flow:

$$\tilde{u} = (1 - y^2, 0,0) \qquad \tilde{p} = - \frac{2x}{R}$$

This is a solution of the Navier Stokes equations in the region $-1 < y < 1$, $-\infty < x,z < \infty$. It is sufficient, according to a theorem of Squire (see Lin's book) to analyze two dimensional disturbances. In this case a stream function $\psi(x,y,t)$ may be introduced; ψ takes the form

$$\psi(x,y,t) = e^{i\alpha(x - ct)} \phi(y) .$$

These eigenfunctions (considered as infinitesimal disturbances) have the form of a wave traveling in the x-direction. The number $i\alpha c$ plays the role of the eigenvalue λ in our earlier analysis. The number c is in general complex, with positive imaginary parts corresponding to growing waves and negative imaginary parts corresponding to decaying disturbances.

The analysis of the linearized stability problem is again quite difficult and has occupied an important place in the subject of hydrodynamic stability since Heisenberg looked at the problem in 1924. It is now well established that the neutral stability curve has the form shown in the diagram below:

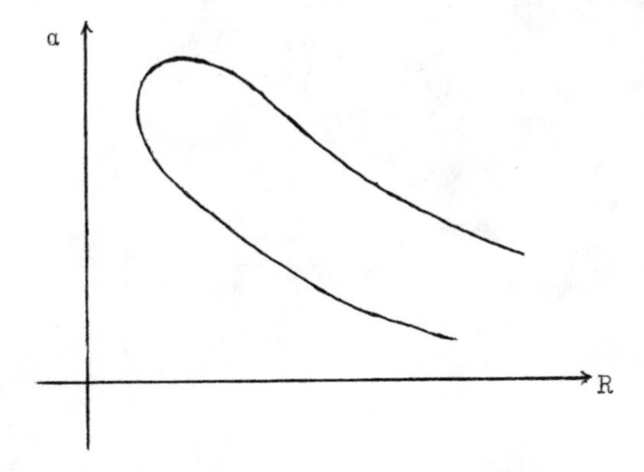

The critical wave speed c is real when the flow loses stability; so $i\alpha c$ is purely imaginary, and this corresponds to the case treated in

Chapter V: loss of stability by a pair of complex conjugate eigenvalues crossing the imaginary axis. A bifurcation analysis which yields traveling wave solutions of the full nonlinear equations has been carried out formally by J. T. Stuart and J. Watson. Numerical work by Reynolds and Potter shows that the bifurcation is subcritical for plane Poiseuille flow.

Bibliography

S. AGMON - Lectures on Elliptic Boundary Value Problems, Van Nostrand, 1965.

S. AGMON, A. DOUGLIS, and L. NIRENBERG - "Estimates Near the Boundary For Solutions of Elliptic Partial Differential Equations Satisfying General Boundary Conditions, I." Comm. Pure Appl. Math. **12**, pp. 623-727 (1959).

H. AMANN - "On the Existence of Positive Solutions of Nonlinear Elliptic Boundary Value Problems." Indiana University Math. Journal **21**, (1971) pp. 125-146.

______ "Existence of Multiple Solutions For Nonlinear Elliptic Boundary Value Problems." Indiana University Math. Journal.

______ "On the Number of Solutions of Nonlinear Equations in Ordered Banach Spaces", Journal Funct. Anal., to appear, 1973.

D. G. ARONSON and L. A. PELETIER - "Stability of Symmetric and Asymmetric Concentrations in Catalyst Particles." Arch. Rat. Mech. Anal. (to appear).

N. BAZLEY and ZWAHLEN - "A Branch of Positive Solutions of Nonlinear Eigenvalue Problems," Manuscripta Mathematica **2**, (1970) pp. 365-374.

M. S. BERGER and P. C. FIFE - "Van Kármán's Equations and the Buckling of a Thin Elastic Plate, II." Comm. Pure Appl. Math. **21** (1968) pp. 227-241.

S. CHANDRASEKHAR - Hydrodynamic and Hydromagnetic Stability. Oxford Press.

______ "The Points of Bifurcation Along the Maclaurin, the Jacobi, and the Jeans Sequences." Astrophysical Journal **137**, p. 1185 (1963).

E. A. CODDINGTON and N. LEVINSON - Theory of Ordinary Differential Equations. NcGraw-Hill, New York, 1955.

D. S. COHEN - "Multiple Stable Solutions of Nonlinear Boundary Value Problems Arising in Chemical Reactor Theory." SIAM Journal on Applied Math. **20** (1971).

D. S. COHEN and H. B. KELLER - "Some Positive Problems Suggested by Nonlinear Heat Generation." J. Math. Mech. **16** (1967) pp. 1361-1376.

D. S. COHEN and R. B. SIMPSON - "Positive Solutions of Nonlinear Elliptic Eigenvalue Problems." J. Math. Mech. **19** (1970), pp. 895-910.

D. COLES - "Transition in Circular Couette Flow." J. Fluid Mech. 21 (1965), pp. 385-425.

R. COURANT and D. HILBERT - Methods of Mathematical Physics, vol. I. Interscience, 1953.

M. G. CRANDALL and P. H. RABINOWITZ - "Bifurcation from Simple Eigenvalues." Jour. Functional Anal. 8 (1971), pp. 321-340.

——— "Nonlinear Sturm-Liouville Eigenvalue Problems and Topological Degree." J. Math. Mech. 19 (1970), pp. 1083-1103.

A. DAVEY, R. C. DI PRIMA, and J. T. STUART - "On the Instability of Taylor Vortices." J. Fluid Mech. 31 (1968), pp. 17-52.

P. FIFE - "The Bénard Problem for General Fluid Dynamical Equations and Remarks on the Boussinesq Equations." Indiana Univ. Math. Jour. 20 (1970), pp. 303-326.

P. FIFE and D. JOSEPH - "Existence of Convective Solutions of the Generalized Bénard Problem." Arch. Rat. Mech. Anal. 33 (1969), pp. 116-138.

D. A. FRANK - KAMENETZKY - Diffusion and Heat Exchange in Chemical Kinetics. Princeton Univ. Press (1955).

K. O. FRIEDRICHS and J. J. STOKER - "The Nonlinear Boundary Value Problem of a Buckled Plate." Amer. Jour. Math. 63 (1941), pp. 839-887.

G. R. GAVALAS - Nonlinear Differential Equations of Chemically Reacting Systems. Springer, New York, 1968.

I. M. GELFAND - "Some Problems in the Theory of Quasilinear Equations." AMS Translations Ser. 2 29, (1963), pp. 295-381.

J. K. HALE - Ordinary Differential Equations. Wiley-Interscience, 1969.

E. HEINZ - "An Elementary Analytic Theory of the Degree of Mapping in n-dimensional Space." Jour. Math. Mech. 8 (1959), pp. 231-247.

E. HOPF - "Abzweigung einer Periodischen Lösung eines Differential Systems." Berichten der Math.-Phys. Klasse der Sächischen Akademie der Wissenschaften zu Leipzig, XCIV, (1942) pp. 1-22.

——— "A Mathematical Example Displaying Features of Turbulence." Comm. Pure Appl. Math. 1 (1948), p. 303.

——— "On Nonlinear Partial Differential Equations." Lecture Series of the Symposium on Partial Differential Equations. Berkeley, 1955. Pub. by University of Kansas.

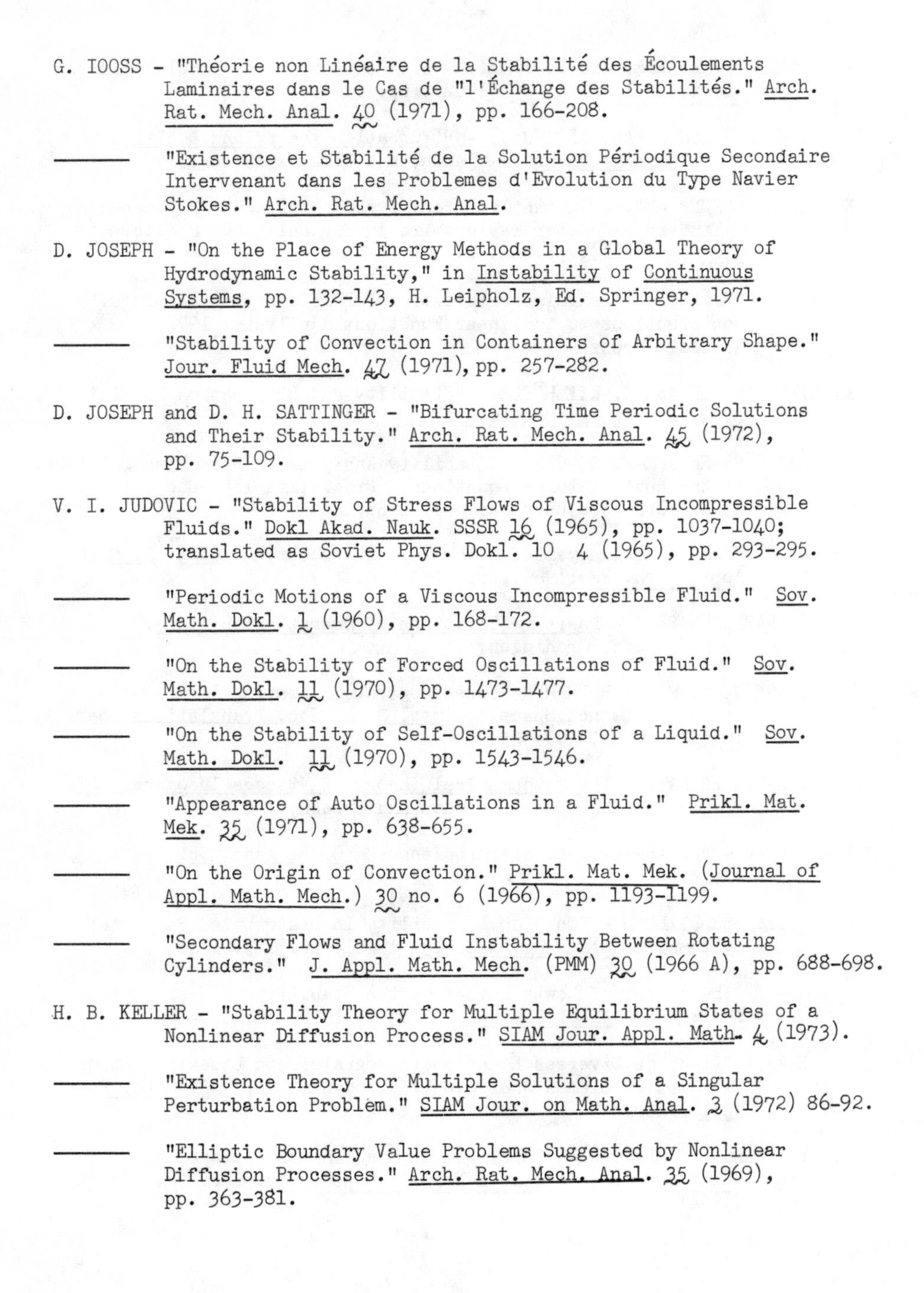

G. IOOSS - "Théorie non Linéaire de la Stabilité des Écoulements Laminaires dans le Cas de "l'Échange des Stabilités." Arch. Rat. Mech. Anal. **40** (1971), pp. 166-208.

——— "Existence et Stabilité de la Solution Périodique Secondaire Intervenant dans les Problemes d'Evolution du Type Navier Stokes." Arch. Rat. Mech. Anal.

D. JOSEPH - "On the Place of Energy Methods in a Global Theory of Hydrodynamic Stability," in Instability of Continuous Systems, pp. 132-143, H. Leipholz, Ed. Springer, 1971.

——— "Stability of Convection in Containers of Arbitrary Shape." Jour. Fluid Mech. **47** (1971), pp. 257-282.

D. JOSEPH and D. H. SATTINGER - "Bifurcating Time Periodic Solutions and Their Stability." Arch. Rat. Mech. Anal. **45** (1972), pp. 75-109.

V. I. JUDOVIC - "Stability of Stress Flows of Viscous Incompressible Fluids." Dokl Akad. Nauk. SSSR **16** (1965), pp. 1037-1040; translated as Soviet Phys. Dokl. 10 4 (1965), pp. 293-295.

——— "Periodic Motions of a Viscous Incompressible Fluid." Sov. Math. Dokl. **1** (1960), pp. 168-172.

——— "On the Stability of Forced Oscillations of Fluid." Sov. Math. Dokl. **11** (1970), pp. 1473-1477.

——— "On the Stability of Self-Oscillations of a Liquid." Sov. Math. Dokl. **11** (1970), pp. 1543-1546.

——— "Appearance of Auto Oscillations in a Fluid." Prikl. Mat. Mek. **35** (1971), pp. 638-655.

——— "On the Origin of Convection." Prikl. Mat. Mek. (Journal of Appl. Math. Mech.) **30** no. 6 (1966), pp. 1193-1199.

——— "Secondary Flows and Fluid Instability Between Rotating Cylinders." J. Appl. Math. Mech. (PMM) **30** (1966 A), pp. 688-698.

H. B. KELLER - "Stability Theory for Multiple Equilibrium States of a Nonlinear Diffusion Process." SIAM Jour. Appl. Math. **4** (1973).

——— "Existence Theory for Multiple Solutions of a Singular Perturbation Problem." SIAM Jour. on Math. Anal. **3** (1972) 86-92.

——— "Elliptic Boundary Value Problems Suggested by Nonlinear Diffusion Processes." Arch. Rat. Mech. Anal. **35** (1969), pp. 363-381.

H. KELLER, J. KELLER, and E. REISS - "Buckled States of Circular Plates." Quart. Appl. Math. 20 (1962), pp. 55-65.

J. B. KELLER and S. ANTMAN (Eds.) - Bifurcation Theory and Nonlinear Eigenvalue Problems. Benjamin, New York, 1969.

K. KIRCHGÄSSNER - "Die Instabilität der Strömung zwischen Rotierenden Zylindern gegenuber Taylor-Wirbeln fur Beliebige Spaltbreiten." ZAMP 12 (1961), pp. 14-30.

——— "Multiple Eigenvalue Bifurcation for Holomorphic Mappings." Contributions to Nonlinear Functional Analysis 1971 Academic Press.

K. KIRCHGÄSSNER and H. KIELHÖFER - "Stability and Bifurcation in Fluid Dynamics." Rocky Mountain J. Math.

K. KIRCHGÄSSNER and P. SORGER - "Stability Analysis of Branching Solutions of the Navier Stokes Equations." Proc. 12th Int. Cong. Appl. Mech., Stanford Univ. August, 1968.

——— "Branching Analysis for the Taylor Problem." Quart. J. Mech. Appl. Math. 22 (1969), pp. 183-210.

M. A. KRASNOSEL'SKII - Positive Solutions of Operator Equations. P. Noordhoff, Groningen.

M. G. KREIN and M. A. RUTMAN - "Linear Operators Leaving Invariant a Cone in a Banach Space." Amer. Math. Soc. Translations, Ser. 1, 10 (1950), pp. 199-325.

O. A. LADYZHENSKAYA - The Mathematical Theory of Viscous Incompressible Flow. 2nd English ed. Gordon and Breach, 1969.

L. LANDAU - "On the Problem of Turbulence." C. R. Acad. Sci. USSR 44, (1944), p. 311.

J. C. LEE and D. LUSS - "On Global Stability in Distributed Parameter Systems." Chem. Eng. Sci. 23 (1968), pp. 1237-1248.

——— "The Effect of Lewis Number on the Stability of a Catalytic Reaction." AIChE Journal, 16, July, 1970, pp. 620-665.

J. LERAY - "Étude de Diverses Équations Intégrales non Linéaires et de Quelques Problèmes que Pose l'Hydrodynamique." J. Math. Pures Appl., série 9, 12 (1933), pp. 1-82.

J. LERAY and J. SCHAUDER - "Topologie et Équations Fonctionelles." Annales Sci. de l'Eclole Normale Superieure, Ser. 3 , 51 (1934), pp. 45-78

N. LEBOVITZ - "On the Fission Theory of Binary Stars." Astrophysical Journal, 174 (1972).

C. C. LIN - The Theory of Hydrodynamic Stability. Cambridge Univ. Press, 1955.

A. LYAPOUNOV - Memoires de l'Academie Imperiale des Sciences de St. Petersburg, VIII, Ser. 22, No. 5.

W. MALKUS and G. VERONIS - "Finite Amplitude Cellular Convection." J. Fluid Mech. 4 (1958) pp. 225-269.

C. B. MORREY - Multiple Integrals in the Calculus of Variations. Springer, New York, 1966.

J. MOSER - "A Rapidly Convergent Iteration Method and Nonlinear Partial Differential Equations." Annali d. Scuola Normale di Pisa, Serie III, Vol. XX (1966) pp. 265-315.

J. R. PARKS - "Criticality Criteria ... as a Function of Activation Energy and Temperature of Assembly." Jour. of Chemical Physics, 34 (1961), pp. 46-50.

A. PAZY and P. H. RABINOWITZ - "A Nonlinear Integral Equation with Applications to Neutron Transport Theory." Arch. Rat. Mech. Anal. 32 (1969), p. 226.

H. POINCARÉ - "Sur l'Équilibre d'une Masse Fluide Animée d'un Mouvement de Rotation." Acta Mathematica 7 (1885), p. 259.

G. PRODI - "Theoremi di Tipo Locale per il Sistema di Navier-Stokes e Stabilita delle Soluzioni Stazionarie." Rend. Sem. Mat. Univ. Padova, 32 (1962), pp. 374-397.

M. H. PROTTER and H. F. WEINBERGER - Maximum Principles in Differential Equations. Prentice-Hall, Englewood Cliff, 1967.

P. H. RABINOWITZ - "Existence and Nonuniqueness of Rectangular Solutions of the Bénard Problem." Arch. Rat. Mech. Anal. 24 (1968) pp. 32-57.

______ "Some Global Results for Nonlinear Eigenvalue Problems." Jour. Functional Analysis, 7 (1971), pp. 487-513.

M. REEKEN - "Stability of Critical Points Under Small Perturbations. Part I." to appear in Manuscripta Mathematica.

______ "Stability of Critical Points Under Small Perturbations. Part II." to appear in Manuscripta Mathematica.

W. C. REYNOLDS and M. C. POTTER - "Finite Amplitude Instability of Parallel Shear Flows." J. Fluid Mech. 14 (1960), p. 336.

RIESZ and NAGY - Functional Analysis. Frederick Ungar.

D. RUELLE and F. TAKENS - "On the Nature of Turbulence." Comm. Math. Phys. **20** (1971), pp. 167-192.

D. SATHER - "Branching of Solutions of Nonlinear Equations in Hilbert Space." Proceedings, Seminar on Bifurcation Theory, Santa Fe, New Mexico, June, 1971.

D. H. SATTINGER - "The Mathematical Problem of Hydrodynamic Stability." Jour. Math. Mech. **19** (1970), pp. 797-817.

______ "Bifurcation of Periodic Solutions of the Navier-Stokes Equations." Arch. Rat. Mech. Anal. **41** (1971), pp. 66-80.

______ "Stability of Bifurcating Solutions by Leray-Schauder Degree." Arch. Rat. Mech. Anal. **43** (1971), pp. 154-166.

______ "Monotone Methods in Nonlinear Elliptic and Parabolic Boundary Value Problems." Indiana Univ. Math. Journal, July, 1972.

J. T. SCHWARTZ - "Nonlinear Functional Analysis." Lecture Notes, N.Y.U. (1963-64).

L. SEGEL - "Nonlinear Hydrodynamic Stability Theory and its Application to Thermal Convection and Curved Flow." in Nonequilibrium Thermodynamics: Variational Techniques and Stability University of Chicago Press, 1966.

J. SERRIN - "Nonlinear Elliptic Equations of Second Order." AMS Symposium in Partial Differential Equations, Berkeley, August, 1971.

______ "On the Stability of Viscous Fluid Motions." Arch. Rat. Mech. Anal. **3** (1959), pp. 1-13.

I. STAKGOLD - "Branching of Solutions of Nonlinear Equations," SIAM Review, **13** (1971), pp. 289-332.

J. T. STUART - "On the Nonlinear Mechanics of Wave Disturbances in Stable and Unstable Parallel Flows." Jour. Fluid Mechanics, **9** (1960), pp. 353-370.

M. M. VAINBERG and V. A. TRENOGIN - "The Methods of Lyapounov and Schmidt in the Theory of Nonlinear Equations and Their Further Development." Russ. Math. Surveys **17** (1962), pp. 1-60.

W. Velte - "Stabilitätsverhalten und Verzweigung Stationäres Lögsungen der Navier-Stokes-schen Gleichungen." Arch. Rat. Mech. Anal. **16** (1964), pp. 97-125.

J. WATSON - "On the Nonlinear Mechanics of Wave Distrubances in Stable and Unstable Parallel Flows, Part II." J. Fluid Mech. **14** (1960), p.336.